MATHS PLUS 6

STAGE 3

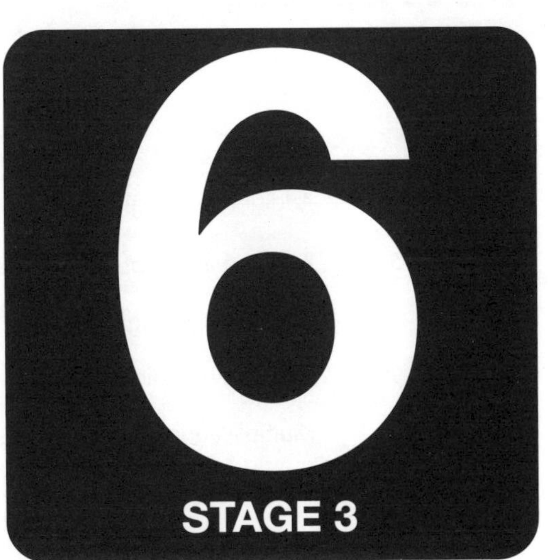

MENTALS AND HOMEWORK BOOK

NEW SOUTH WALES SYLLABUS

Harry O'Brien
Greg Purcell

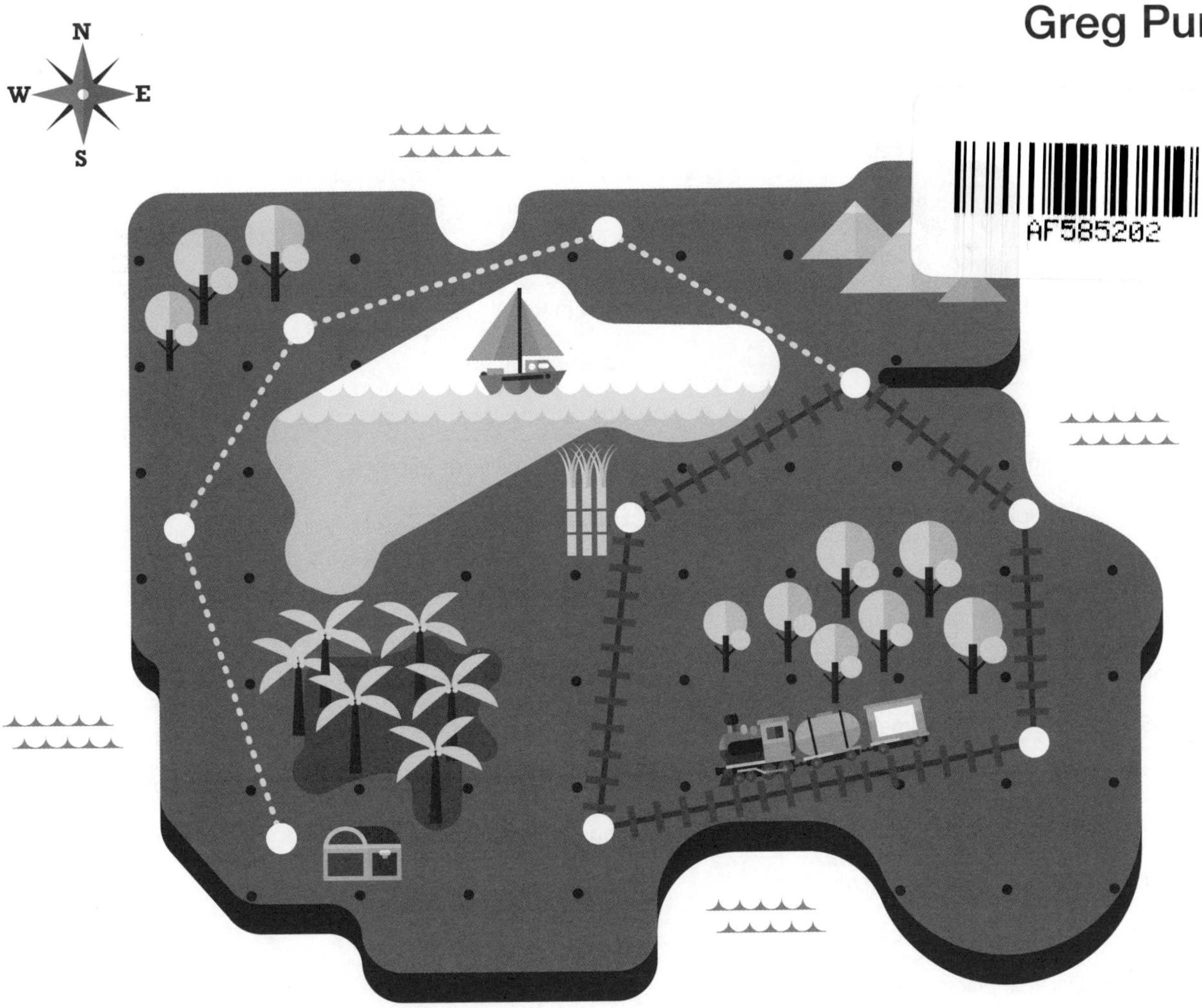

OXFORD
UNIVERSITY PRESS

Contents

Contents

NSW Syllabus Outcomes

Units	1	2	3	4
NUMBER AND ALGEBRA				
MA3-RN-01				
Applies an understanding of place value and the role of zero to represent the properties of numbers				
MA3-RN-02				
Compares and orders decimals up to 3 decimal places				
MA3-RN-03				
Determines percentages of quantities, and finds equivalent fractions and decimals for benchmark percentage values				
MA3-AR-01				
Selects and applies appropriate strategies to solve addition and subtraction problems				
MA3-MR-01				
Selects and applies appropriate strategies to solve multiplication and division problems				
MA3-MR-02				
Constructs and completes number sentences involving multiplicative relations, applying the order of operations to calculations				
MA3-RQF-01				
Compares and orders fractions with denominators of 2, 3, 4, 5, 6, 8 and 10				
MA3-RQF-02				
Determines $\frac{1}{2}$, $\frac{1}{4}$, $\frac{1}{5}$ and $\frac{1}{10}$ of measures and quantities				
MEASUREMENT AND SPACE				
MA3-GM-01				
Locates and describes points on a coordinate plane				
MA3-GM-02				
Selects and uses the appropriate unit and device to measure lengths and distances including perimeters				
MA3-GM-03				
Measures and constructs angles, and identifies the relationships between angles on a straight line and angles at a point				
MA3-2DS-01				
Investigates and classifies two-dimensional shapes, including triangles and quadrilaterals based on their properties				
MA3-2DS-02				
Selects and uses the appropriate unit to calculate areas, including areas of rectangles				
MA3-2DS-03				
Combines, splits and rearranges shapes to determine the area of parallelograms and triangles				
MA3-3DS-01				
Visualises, sketches and constructs three-dimensional objects, including prisms and pyramids, making connections to two-dimensional representations				
MA3-3DS-02				
Selects and uses the appropriate unit to estimate, measure and calculate volumes and capacities				
MA3-NSM-01				
Selects and uses the appropriate unit and device to measure the masses of objects				
MA3-NSM-02				
Measures and compares duration, using 12- and 24-hour time and am and pm notation				
STATISTICS AND PROBABILITY				
MA3-DATA-01				
Constructs graphs using many-to-one scales				
MA3-DATA-02				
Interprets data displays, including timelines and line graphs				
MA3-CHAN-01				
Conducts chance experiments and quantifies the probability				
MAO-WM-01 Working mathematically				
Develops understanding and fluency in mathematics through exploring and connecting mathematical concepts, choosing and applying mathematical techniques to solve problems, and communicating their thinking and reasoning coherently and clearly				

	5	6	7	8	9	10	11	12	13	14	15	16	17	18	19	20	21	22	23	24	25	26	27	28	29	30	31	32	33	34	35
NUMBER AND ALGEBRA																															
MA3-RN-01																															
MA3-RN-02																															
MA3-RN-03																															
MA3-AR-01																															
MA3-MR-01																															
MA3-MR-02																															
MA3-RQF-01																															
MA3-RQF-02																															
MEASUREMENT AND SPACE																															
MA3-GM-01																															
MA3-GM-02																															
MA3-GM-03																															
MA3-2DS-01																															
MA3-2DS-02																															
MA3-2DS-03																															
MA3-3DS-01																															
MA3-3DS-02																															
MA3-NSM-01																															
MA3-NSM-02																															
STATISTICS AND PROBABILITY																															
MA3-DATA-01																															
MA3-DATA-02																															
MA3-CHAN-01																															
MAO-WM-01 Working mathematically																															

UNIT 1

Number and Algebra

SET 1 Basic

1 24 + 8

2 8 × 6

3 34 – 17

4 8 ☐ 4 = 32

5 27 ☐ 9 = 3

6 9^2

7 Divide 72 by 8.

8 12 ☐ 4 = 48

9 Product of 11 and 5

10 Sum of 9, 5 and 10

11 Quotient of 54 and 6

12 Cents in $8.69

13 (6 + 2) × (3 + 6)

14 Factors of 45

15 I had $20 but I spent $13.50. How much have I left?

$ ☐

SET 2 Addition of 4-digit numbers

1 $\begin{array}{r} 3704 \\ +\ 2204 \\ \hline \end{array}$

2 $\begin{array}{r} 3628 \\ +\ 4719 \\ \hline \end{array}$

3 $\begin{array}{r} 3876 \\ +\ 666 \\ \hline \end{array}$

4 $\begin{array}{r} 4720 \\ +\ 7089 \\ \hline \end{array}$

5 $\begin{array}{r} 3625 \\ 703 \\ 80 \\ +\ 924 \\ \hline \end{array}$

6 $\begin{array}{r} 3078 \\ 609 \\ 860 \\ +\ 7436 \\ \hline \end{array}$

7 Michael saved $1206 in February, $986 in March and $1807 in April. What was his total for the 3 months?

8 How much is Stella's set of 4 paintings worth if they were valued at $4125, $3250, $1899 and $5455?

Space Measuring angles

Measure each angle using a protractor.

1

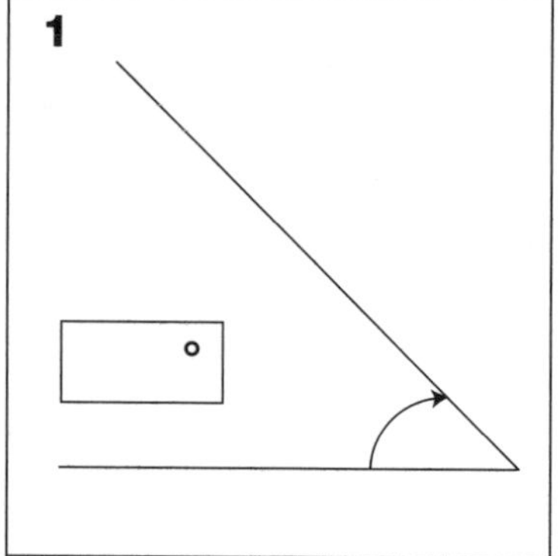

2

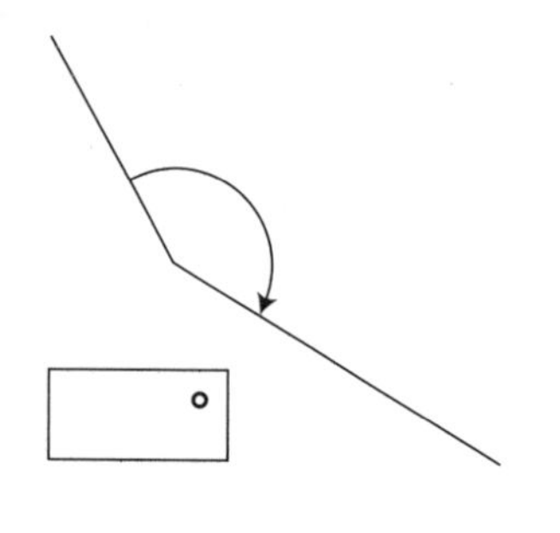

3

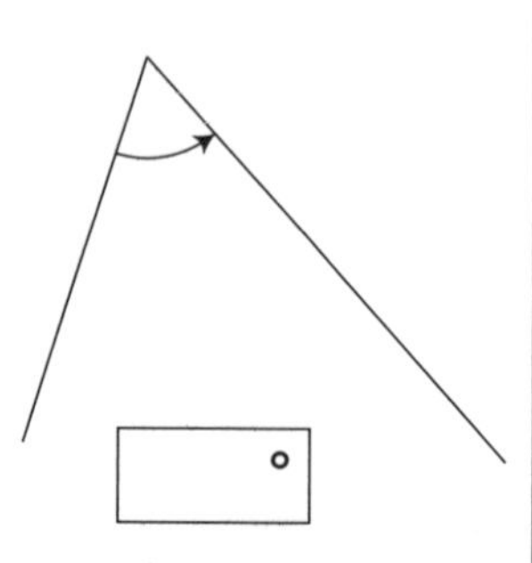

4

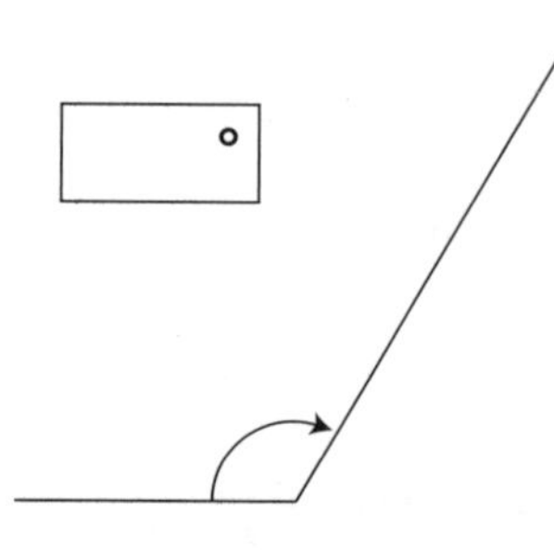

Number and Algebra

SET 3 Multiplication strategies

Solve the facts.

1

	×5
3	
5	
7	
8	
6	

2

	×7
2	
5	
7	
9	
8	

3

	×9
2	
5	
7	
9	
8	

To multiply by 5, multiply by 10 then halve your answer.

4 42 × 5 =

5 64 × 5 =

6 48 × 5 =

7 84 × 5 =

8 80 × 5 =

9 18 × 5 =

To multiply by 4, use the double and double again strategy.

10 13 × 4 =

11 26 × 4 =

12 42 × 4 =

13 18 × 4 =

To multiply by 8, use the double, double and double again strategy.

14 13 × 8 =

15 21 × 8 =

16 32 × 8 =

17 15 × 8 =

SET 4 Extension

1 5.8 + 3.6

2 20 × 5 × 8

3 Value of 8 in 78 321

4 $\frac{4}{10}$ of 50

5 Round 8.3 to the nearest whole number.

6 Add 1.7 to 4.4.

7 5 tickets at $1.65 each

8 Subtract the sum of 50 and 40 from 200.

9 Add the product of 6 and 8 to 42.

10 Round 1561 to the nearest thousand.

11 $\frac{7}{100}$ is less than 0.5. True or false?

12 If the water in the jug is 87°C, how many more degrees before it boils?

13 If Hugo ran 0.8 km each day, how far did he run in a school week?

14 Round and estimate 399 × 19.

15 If I spent $2, $1 and 65c, how much change would I receive from $5?

16 How many 250 g bags of rice can be made from a $3\frac{3}{4}$ kg bag?

Measurement Kilometres

Each band on the radar is equal to 10 km. Estimate the distances of the submarines from the centre of the radar screen.

1 Submarine A [ ] km

2 Submarine B [ ] km

3 Submarine C [ ] km

4 Submarine D [ ] km

5 Submarine E [ ] km

6 Submarine F  km

HMAS *Pinafore* radar

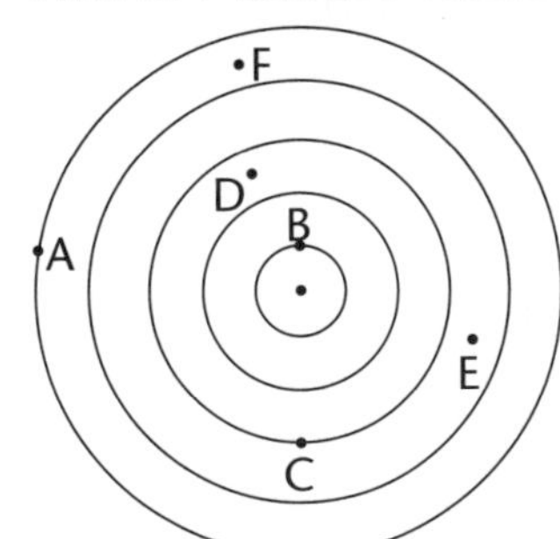

Number and Algebra

SET 1 Basic

1 ☐ − 8 = 3

2 85 × 0

3 21 ÷ 3

4 400 − 150

5 7^2

6 Date after 11 February

7 49 ÷ 7

8 5c × 100

9 Difference between 19 and 90

10 Season before spring

11 28 ☐ 7 = 4

12 Quotient of 24 and 6

13 How many quarters in 5?

14 Value of 7 in 17 940

15

SET 2 Addition and subtraction strategies

Add these numbers mentally.

1 37 + 49 =

2 68 + 75 =

3 81 + 67 =

4 99 + 36 =

5 121 + 56 =

6 379 + 85 =

7 741 + 126 =

8 899 + 257 =

Subtract these numbers mentally.

9 76 − 29 =

10 89 − 57 =

11 68 − 34 =

12 149 − 37 =

13 259 − 58 =

14 741 − 127 =

Working Mathematically

Round each number to the nearest 100 to estimate an approximate answer. The first one has been done for you.

	Question	Rounded to 100	Approximate answer
15	395 + 206	400 + 200	600
16	591 − 298		
17	513 + 387		
18	785 − 589		
19	372 + 329		
20	882 − 286		

Space Two-dimensional shapes

Sketch a shape to match each description.

1 I am a shape with 5 sides the same length and 5 identical angles.	2 I am a shape that has 3 sides and 1 right angle.	3 I am a shape that has 4 sides the same length but 2 sets of angles.	4 I am a shape with 8 sides of equal length and 8 identical angles.

Number and Algebra

SET 3 Fractions, decimals and percentages

	Fraction	Decimal	%
1	$\frac{10}{100}$		
2	$\frac{25}{100}$		
3	$\frac{7}{10}$		
4	$\frac{20}{100}$		
5	$\frac{1}{2}$		
6	$\frac{1}{4}$		
7	$\frac{3}{4}$		

Mathematical Reasoning

Order from smallest to largest.

8	$\frac{27}{100}$	30%	0.29	
9	35%	$\frac{53}{100}$	0.33	
10	$\frac{99}{100}$	9%	0.9	
11	0.54	$\frac{1}{2}$	49%	
12	4%	$\frac{3}{10}$	0.21	
13	$\frac{9}{100}$	90%	0.95	
14	$\frac{70}{100}$	7%	0.03	

SET 4 Extension

1 6×15

2 $\frac{23}{100} = \square$ %

3 Value of 6 in 37 206

4 Average of 45, 25 and 35

5 $\frac{3}{4}$ of $96

6 How many faces has a rectangular prism?

7 Round 63 209 to the nearest 100.

8 If 3 kg cost $18.60, how much would 5 kg cost?

9 $\frac{38}{100}$ = 38%. True or false?

10 $300 \div 2 \times 10 - 750 + 53$

Working Mathematically

Solve these missing number sentences. Each shape has the same value in every question. For example, the square is always equal to 9.

11 $\square \times \triangle = 54$

12 ⬡ $\times 2 = \triangle$

13 $\square \div$ ⬡ $=$ ⬡

14 $\square \times$ ⬡ $+ \triangle = 33$

15 $72 \div \square \times \triangle = 48$

16 $54 \div \triangle =$ ⬡ $\times$ ⬡

Measurement Square centimetres

Calculate the area of these rectangles in square centimetres (cm^2).

1 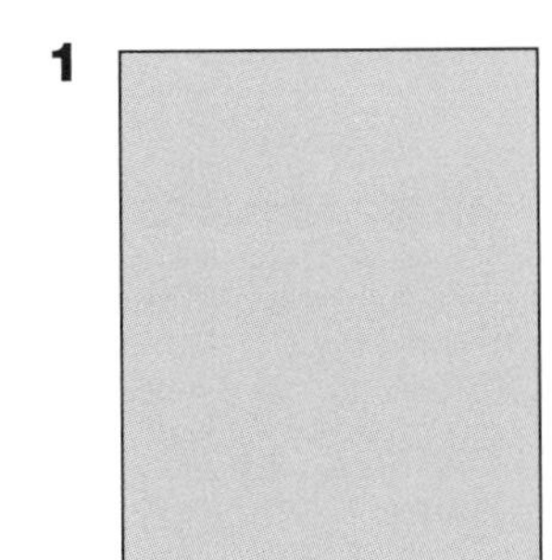

2 

3 

	Length × Width = Area		
1			
2			
3			

Number and Algebra

SET 1 Basic

1 8 + 7

2 3 × 4

3 13 – 4

4 6 ☐ 3 = 18

5 18 ☐ 6 = 12

6 3^2

7 Divide 12 by 2.

8 18 ☐ 3 = 6

9 Product of 7 and 4

10 Sum of 7 and 14

11 Difference between 14 and 21

12 (3 + 7) × 7

13 Is 28 a multiple of 7?

14 1207, 1210, 1213, ☐

15

Peter has run 550 m of an 800 m race. How much further does he need to run to finish the race?

☐ m

SET 2 Subtracting 4-digit numbers

1 $\begin{array}{r} 5000 \\ -\ 3000 \\ \hline \end{array}$

2 $\begin{array}{r} 8000 \\ -\ 4000 \\ \hline \end{array}$

3 $\begin{array}{r} 5425 \\ -\ 3000 \\ \hline \end{array}$

4 $\begin{array}{r} 8960 \\ -\ 4000 \\ \hline \end{array}$

5 $\begin{array}{r} 6982 \\ -\ 3421 \\ \hline \end{array}$

6 $\begin{array}{r} 8536 \\ -\ 3425 \\ \hline \end{array}$

7 $\begin{array}{r} 8516 \\ -\ 3344 \\ \hline \end{array}$

8 $\begin{array}{r} 8643 \\ -\ 4461 \\ \hline \end{array}$

9 $\begin{array}{r} 5941 \\ -\ 3799 \\ \hline \end{array}$

10 $\begin{array}{r} 6381 \\ -\ 1295 \\ \hline \end{array}$

11 $\begin{array}{r} 8161 \\ -\ 3429 \\ \hline \end{array}$

12 $\begin{array}{r} 7564 \\ -\ 3818 \\ \hline \end{array}$

Statistics and Probability Picture graphs

Hair colours Year 6

Red Brown Black Fair Blond

KEY 1 face = 4 children

Use the key to answer the questions.

How many children had:

1 red hair? ________

2 brown hair? ________

3 black hair? ________

4 fair hair? ________

5 blond hair? ________

Number and Algebra

SET 3 Revising 3-digit division

1 $3\overline{)969}$

2 $4\overline{)568}$

3 $5\overline{)175}$

4 $6\overline{)156}$

5 $7\overline{)763}$

6 $8\overline{)744}$

7 Share $64 among 8.

8 How many fives in 90?

9 $450 \div 10$

10 What is the quotient when 357 is divided by 7?

11 Share 424 blocks among 4 groups.

12 Lisa's mum won $924 in the lottery. If she shared it with another 6 people, how much did each person receive?

13 540 spectators were seated in 10 rows. How many in each row?

14 A group of 217 boys was divided into 7 teams. How many in each team?

SET 4 Extension

1 Value of 7 in 73 256

2 Write $\frac{8}{100}$ as a percentage.

3 1.8 m = ☐ cm

4 Round 3.9 to the nearest whole number.

5 Average of 43, 36 and 50

6 3113, 3128, 3143, ☐

7 $9^2 - 5$

8 Would a hot summer's day be 20°C, 35°C or 100°C?

9 How many grams in 7.3 kg?

10 Round 29 306 to the nearest 1000.

11 $\frac{3}{10}$ of 200

12 Perimeter of a pentagon with sides of 1.5 cm

13 How many axes of symmetry has a square?

14 $\frac{3}{5} = \frac{\square}{40}$

15 Tom ate 25% of the cake. Jill ate 0.27 of it and Max ate 33 hundredths of it. How much was left?

16 How much does 9 kg of meat cost at $3.50 kg?

Space Constructing angles

Use a protractor to create these angles.

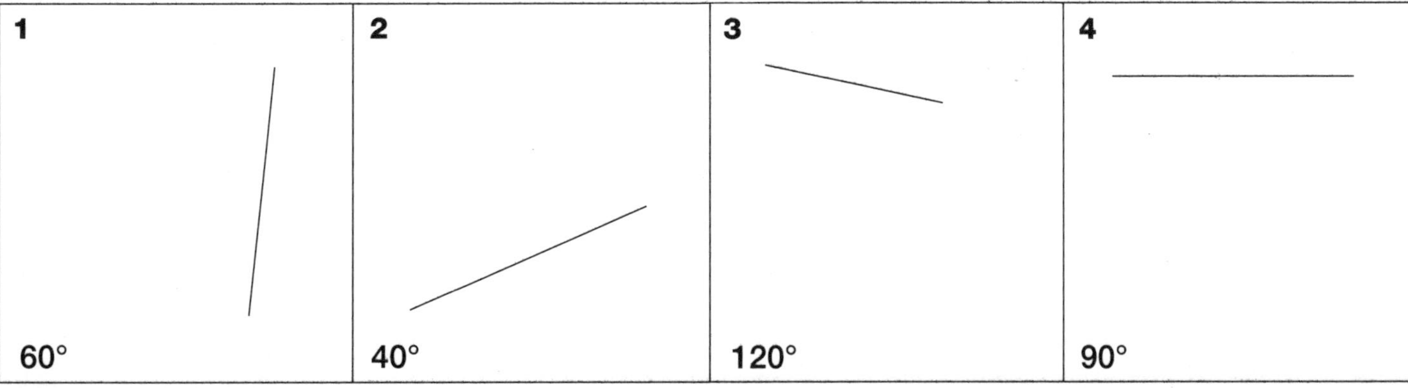

Number and Algebra

SET 1 Basic

1 9×5
2 $16 - 9$
3 $14 + 8$
4 4^2
5 16 ☐ 4 = 12
6 18 ☐ 3 = 6
7 Divide 35 by 7
8 20 ☐ 20 = 40
9 $\frac{1}{4} \times \$28$
10 Quotient of 25 and 5
11 Factors of 9
12 Are 18 and 24 multiples of 6?
13 Is 21 a prime number?
14 How many cm in 7 m?
15
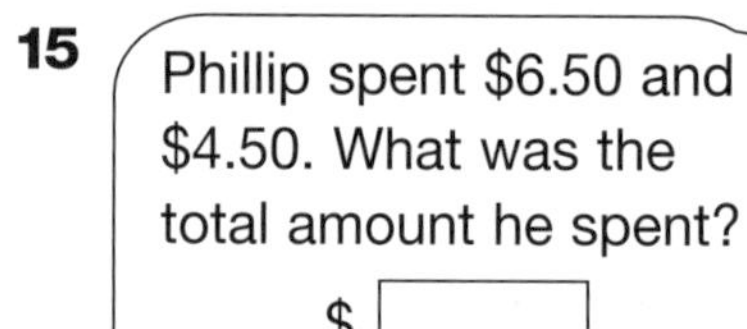

SET 2 Addition of 4-digit numbers

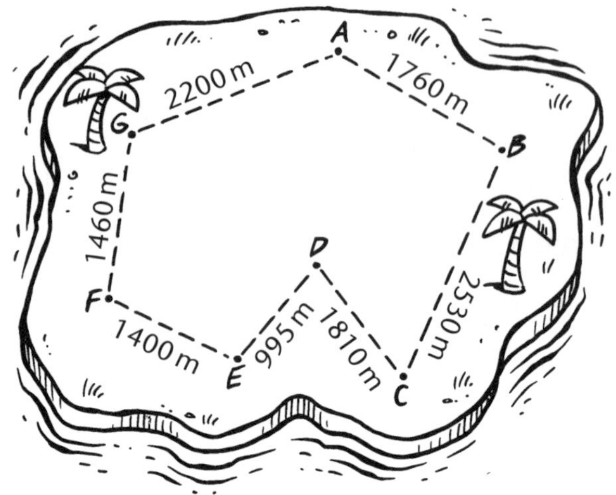

Bill's Bus Service charges customers $2 for every kilometre they travel. Calculate the distance and the cost for each journey.

	Journey	Metres	Cost
1	A to C via B		
2	B to E via C and D		
3	F to B via G and A		
4	E to A via F and G		
5	G to D via A, B and C		
6	B to F via C, D and E		
7	F to C via G, A and B		

Number and Algebra Negative numbers

Calculate the answers. The number lines may help you.

1 $-2 + 4 + 3 - 6 =$ ☐

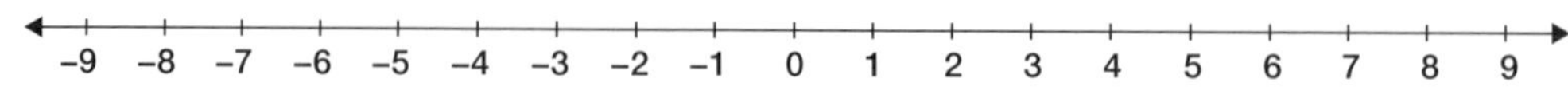

2 $-8 + 3 + 5 - 7 =$ ☐

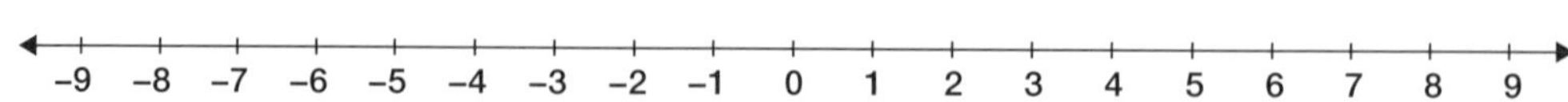

3 $-6 + 7 - 9 + 2 =$ ☐

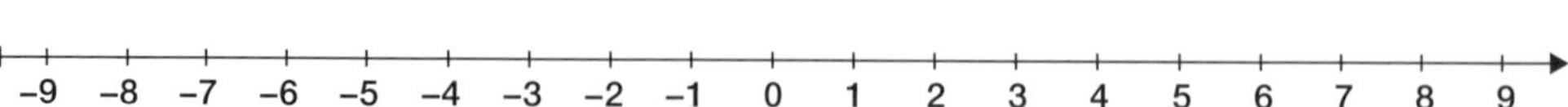

Number and Algebra

SET 3 Place value

Draw beads on the abacuses to represent the numbers.

1 257 379

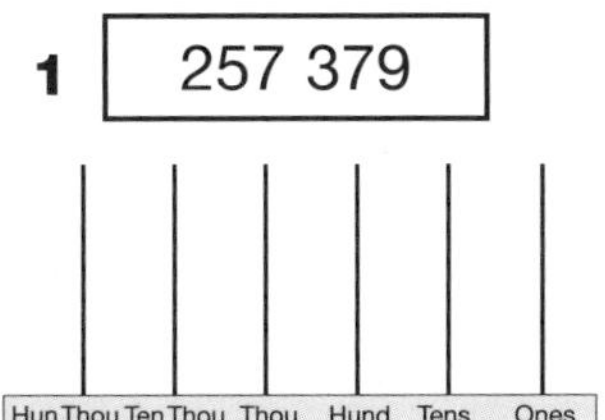

2 703 825

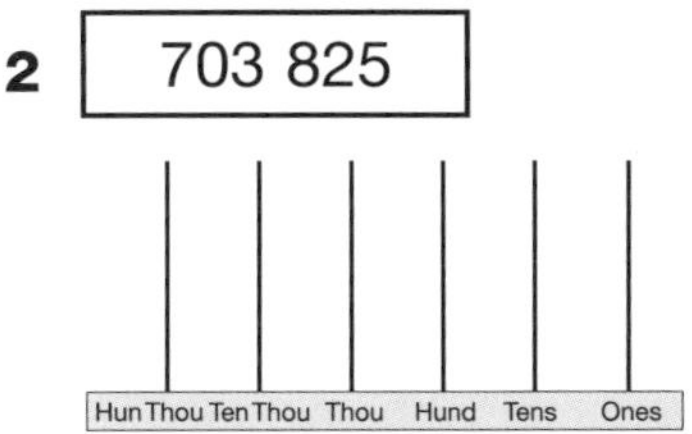

3 Write the number sixty-four thousand, nine hundred and twenty-eight.

4 What are the next 2 numbers after 52 817?

5 How many whole dollars in 19 361 cents?

6 Round 46 985 to the nearest thousand.

7 Subtract 1000 from 158 695.

8 What number is ten thousand less than 262 340?

9 How many hundreds are in 34 972?

10 What number is one thousand more than 442 186?

11 What are the 2 numbers before 214 901?

12 What is the next odd number after 25 131?

13 What number is ten thousand less than 985 970?

SET 4 Extension

1 160 – 85

2 $\frac{3}{4} = \frac{\square}{12}$

3 Value of 7 in 86 791

4 Complete this sequence:
5404, 5504, 5604, ☐, ☐

5 How many months in half a year?

6 $\frac{3}{4} \times 48 + 207$

7 Round $18.61 to the nearest dollar.

8 Average of 45, 26, 39 and 450

9 180 seconds = ☐ minutes

10 How much is 4 kg at $1.70 a kg?

11 1 km ÷ 4 = ☐ m

12 Place in a sequence: $\frac{1}{2}, \frac{1}{4}, \frac{1}{3}, \frac{3}{4}$

13 9:25 + 58 minutes

14 $0.72 × 9

15 How much change would I receive from $20 if I bought three $1.85 ice creams?

16 If 3 kg cost $1.95, how much would 10 kg cost?

17 Round to estimate an answer to 51 × 494.

Measurement and Space Classifying three-dimensional objects

Complete the grid to classify the shapes.

1

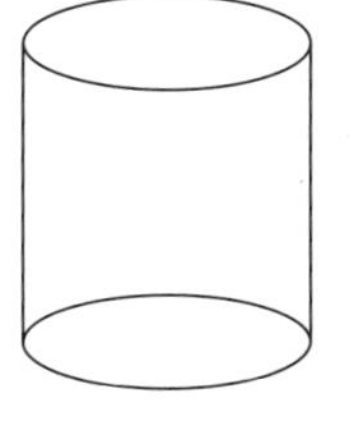

2

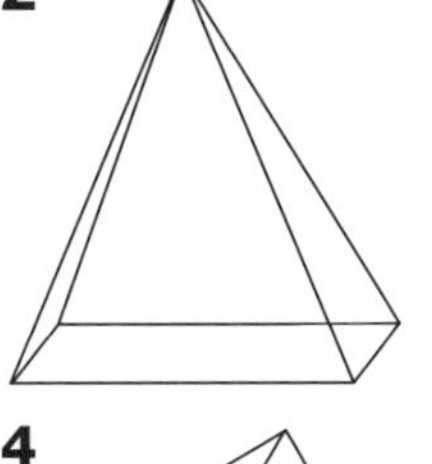

3

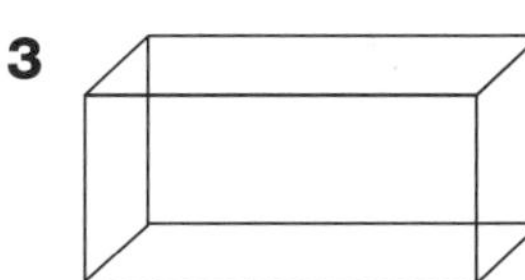

4

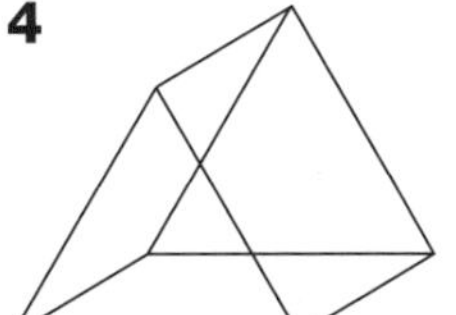

	Name	Faces	Vertices	Edges
1				
2				
3				
4				

Number and Algebra

SET 1 Basic

1 18 + 6

2 4 × 9

3 25 – 16

4 5 ☐ 8 = 40

5 9 ☐ 3 = 3

6 8^2

7 Divide 64 by 8.

8 11 ☐ 3 = 33

9 Product of 10 and 6

10 Sum of 8, 6 and 9

11 Quotient of 54 and 9

12 Cents in $9.61

13 (4 + 5) × (4 + 6)

14 Factors of 27

15

What are the next 2 numbers in this sequence?
27, 54, 108, ☐, ☐

SET 2 Multiplication

1 7 × 5

2 7 × 50

3 7 × 500

4 98 × 10

5 98 × 100

6 98 × 1000

7 Round to estimate an answer to 395 × 5.

8 Multiply 603 by 7.

9 Product of 417 and 3

10 How much are 9 tins of paint at $7.35 each?

Complete these algorithms.

11 $\begin{array}{r} 263 \\ \times\ \ \ 4 \\ \hline \end{array}$ 12 $\begin{array}{r} 589 \\ \times\ \ \ 3 \\ \hline \end{array}$ 13 $\begin{array}{r} 407 \\ \times\ \ \ 6 \\ \hline \end{array}$

14 $\begin{array}{r} 872 \\ \times\ \ \ 3 \\ \hline \end{array}$ 15 $\begin{array}{r} 483 \\ \times\ \ \ 6 \\ \hline \end{array}$ 16 $\begin{array}{r} 338 \\ \times\ \ \ 7 \\ \hline \end{array}$

Statistics and Probability Chance predictions

Matt dropped his bag of 30 marbles and 6 rolled out. Use the marbles that rolled out to estimate how many marbles of each colour might be in the bag.

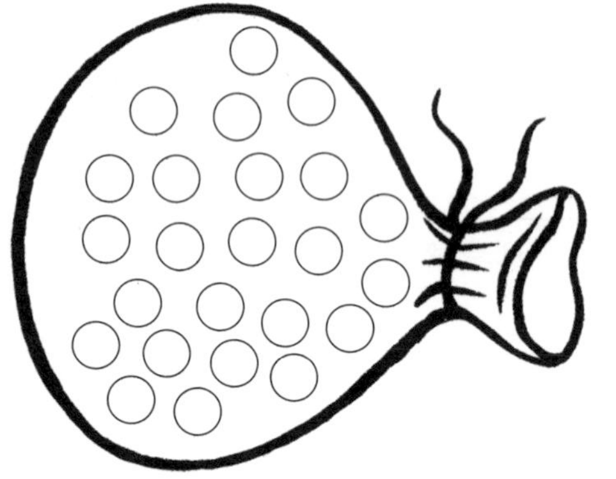

1 Red

2 Blue

3 Green

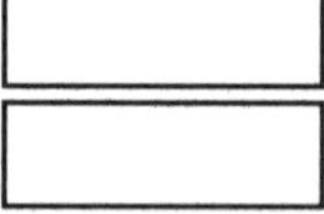

Number and Algebra

SET 3 Decimals, percentages and fractions

Shade the grids to display the fractions, then write the equivalent percentage and decimal.

1 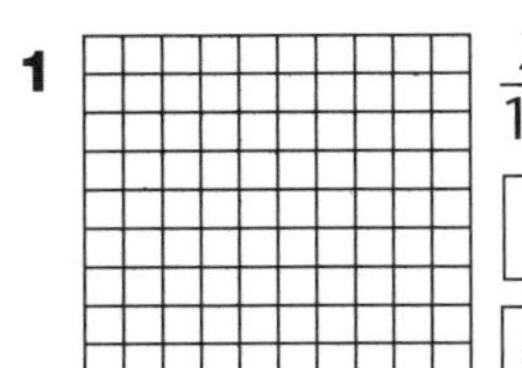 $\frac{29}{100}$ 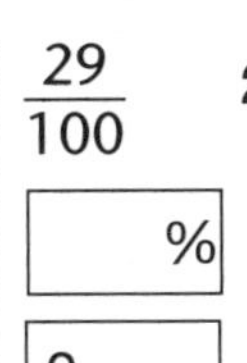

2 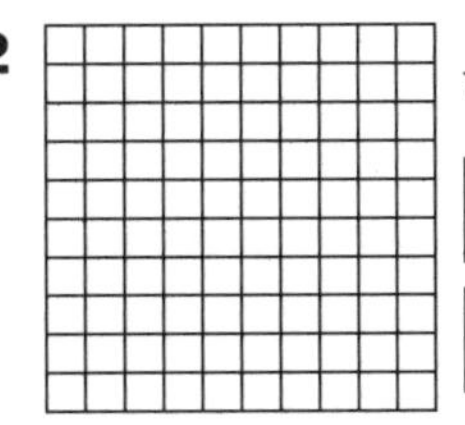 $\frac{37}{100}$ 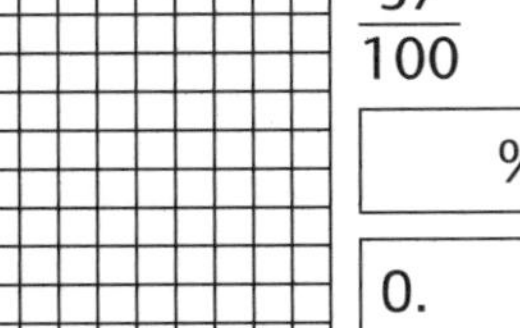

Write decimals for each fraction.

3 $\frac{1}{10}$ =

4 $\frac{5}{10}$ =

5 $\frac{5}{100}$ =

6 $\frac{3}{4}$ =

7 $\frac{1}{2}$ =

8 $\frac{27}{100}$ =

9 $\frac{9}{100}$ =

10 $\frac{97}{100}$ =

Express each decimal as a percentage.

11 0.7 =

12 0.25 =

13 0.99 =

14 0.05 =

15 0.35 =

16 0.07 =

17 0.09 =

18 0.75 =

19 Order these decimals and fractions from smallest to largest.

0.35	$\frac{38}{100}$	70%	0.09

SET 4 Extension

1 440 – 92

2 $\frac{3}{4} + \frac{3}{4} + \frac{1}{4}$

3 Value of 6 in 96 401

4 Complete this sequence:
6924, 6914, 6904, ____, ____

5 At what temperature does water begin to boil?

6 How many dollars in 19 615 cents?

7 Order $\frac{3}{5}$, $\frac{1}{2}$, 0.47 and 40%.

8 6 hours = ____ minutes

9 How much change would I receive from \$50 if I spent \$3.28?

10 How far did we travel if we drove for $4\frac{1}{2}$ hours at an average speed of 68 km/h?

11 What is the third angle of a triangle if the other 2 are 27° and 58°?

12 How many m^2 in one hectare?

13 $5 \times 9 + 2 \times 9 + 3 \times 9$

14 $\frac{3}{5}$ of 200 ÷ $\frac{3}{4}$ of 16

15 How much are 13 brushes at \$3.75 each?

16 What is the perimeter of an octagon with 115 mm sides?

Measurement Grams

1 How many of each item could fit in the box?

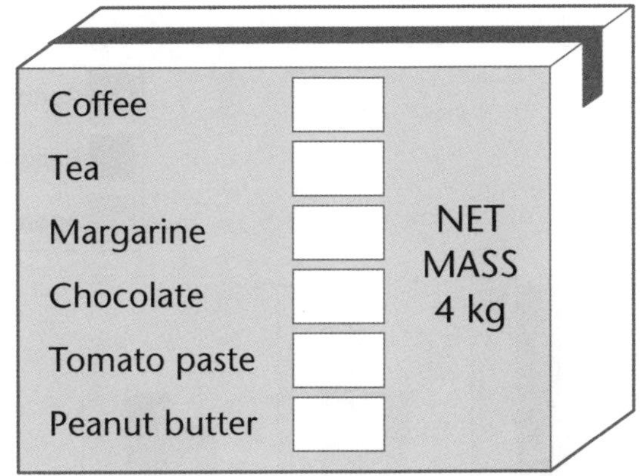

2 Mr Bean bought 4 items with a total mass of 1275 grams. What were the items?

Number and Algebra

SET 1 Basic

1 100 ☐ 25 = 75

2 8 × 3

3 13 – 6

4 24 ÷ 4

5 80 + 80

6 100 ☐ 120 = 220

7 100 ☐ 2 = 50

8 Product of 7 and 9

9 Difference between 24 and 6

10 Value of 6 in 6135

11 Factors of 20

12 1405, 1411, 1417, ☐

13 Is 9 a prime number?

14 $3\frac{1}{2}$ min = ☐ seconds

15

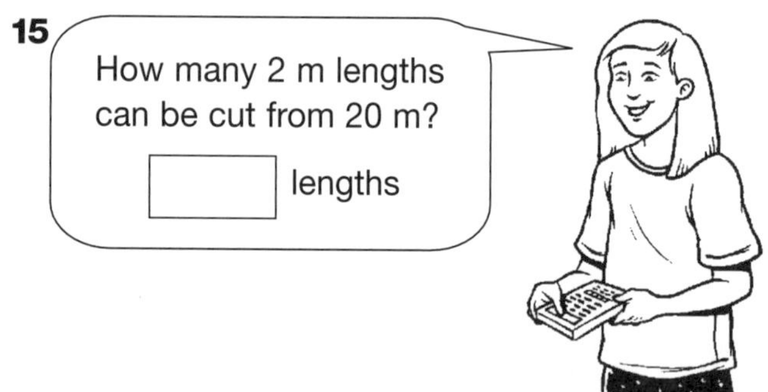

SET 2 4-digit division

Solve these problems.

1 I paid $5250 for 5 televisions. How much were they each? 

2 A car travelled 1835 km in 5 days. What was the average distance travelled each day? 

3 6930 bags were packed into 10 boxes. How many bags in each box? 

4 John had a bag of 981 stamps to share among himself and 2 friends. How many stamps did each person receive? 

5 Melissa earned $2352 over 4 weeks. What were her average weekly earnings? 

6 1000 mL of perfume was poured into 8 mL sample bottles. If each bottle was sold for $5, what was their total value? 

Statistics and Probability Side-by-side column graphs

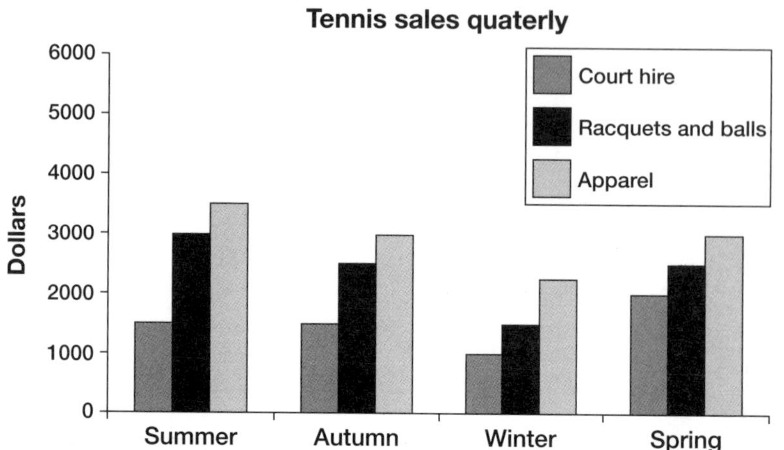

Which season was best for:

1 Apparel sales? ____________

2 Racquet and ball sales? ____________

3 Court hire? ____________

4 How much was the total sales for apparel during winter? ____________

Number and Algebra

SET 3 Number patterns

Complete the number sequences.

1

1	2	3	4	5	6	7	8
4		12			24		

2

12	14	16	18	20	22	24	26
5				13			19

3

1	2	3	4	5	6	7	8
11			44				88

Complete the sequence then write a rule.

4

12	15	18	21	24	27	30	33
		6					11

Rule: ______________________________

Working Mathematically

5 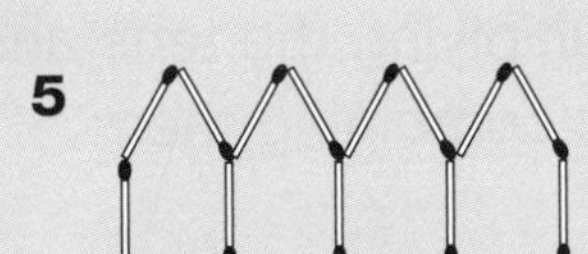

If this pattern of matchstick houses was continued, how many could be built with 25 matchsticks?

______________ houses

SET 4 Extension

1 7×15

2 If 3 kg costs \$3.54, how much would 6 kg cost?

3 Value of 4 in 1 400 962

4 Average of 41, 52 and 18

5 Order 35%, 0.2, $\frac{1}{4}$ and 0.31.

6 How much is 3.5 kg at \$8 per kg?

7 Round \$19.25 to the nearest dollar.

8 2 km ÷ 4

9 $(16 - 12) \times 5$

10 $280 + 40 \div 8$

11 How much change would I receive from \$50 if I spent \$7.18?

12 $\frac{7}{10}$ of a kilometre

13 $\frac{1}{10} + \frac{17}{100} + \frac{9}{100} =$

14 $\frac{3}{8} + \frac{7}{8} + \frac{5}{8} =$

15 Round 9.32 to the nearest tenth.

16 $\frac{3}{5} \times 75 + \frac{2}{5} \times 50$

Working Mathematically

Complete the decimal addition chart.

	+	7.431	5.323	2.249
17	0.05			
18	0.003			
19	0.5			
20	0.001			

Measurement Time

Draw the analog time for each clock.

1 2:30 am

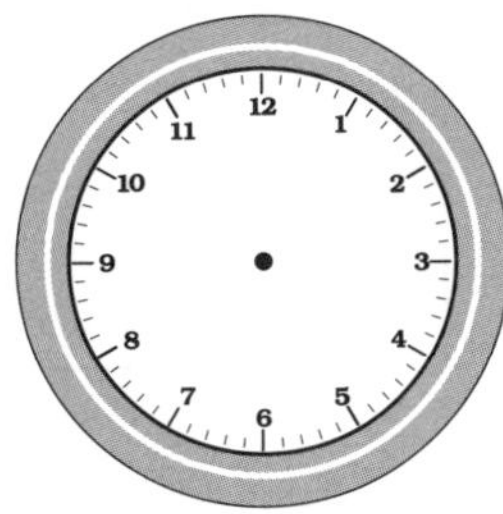

2 7:15 am

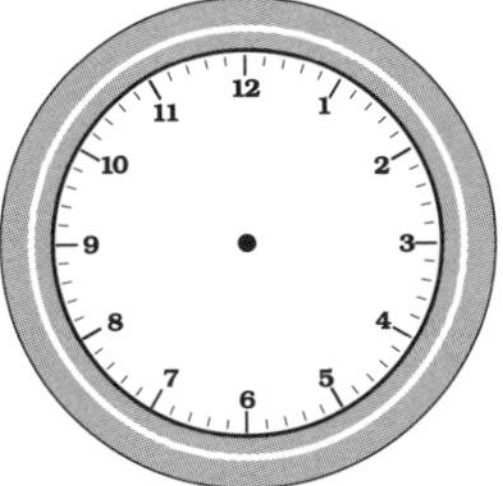

3 4:45 pm

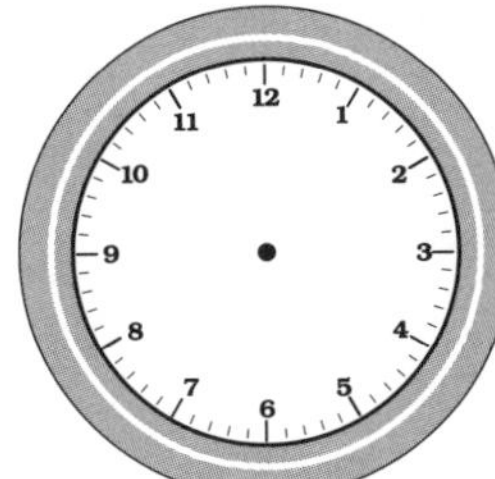

4 7:25 pm

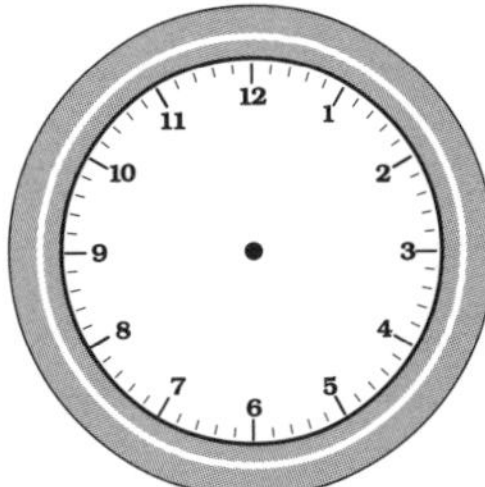

5 10:23 pm

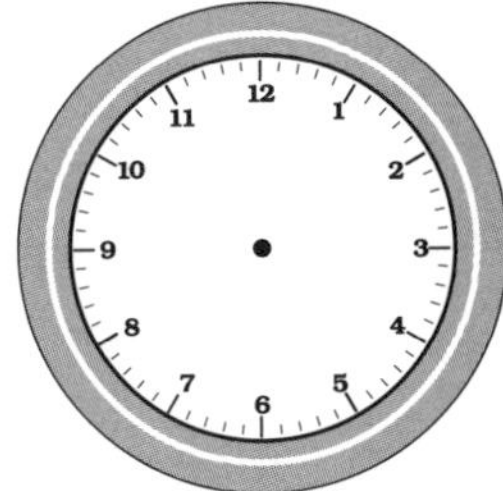

UNIT 7

Number and Algebra

SET 1 Basic

1 $48 \div 6$
2 $(4 + 3) \times 5$
3 $4 + 4 + 4 + 4$
4 7^2
5 Product of 9 and 2
6 Cents in $4.60
7 $(4 + 3) \times (5 + 1)$
8 $3c \times 100$
9 56 ☐ 8 = 7
10 Divide 90 by 10.
11 Season after summer
12 Is 29 a multiple of 7?
13 Quotient of 63 and 7
14 $(6 + 2) \times 4$
15

How do you write one hundred and ninety-two in Hindu–Arabic numerals? ☐

SET 2 Multiplication strategies

Calculate mentally.

1 3×9
2 3×90
3 3×900
4 30×90
5 4×60
6 9×80
7 7×800
8 30×50
9 60×70
10 50×90

Estimate the answers by rounding the larger number in each multiplication.

11 29×3
12 49×4
13 57×6
14 99×5
15 71×5
16 39×20
17 52×30
18 199×20

Calculate mentally by multiplying the tens and the ones separately, then combining them.

19 35×4
20 42×5
21 55×6
22 44×4
23 27×8

Statistics and Probability Frequency/probability

The frequency chart shows the number of times different colour marbles were picked out of a bag.

Red	Blue	Green	Gold
~~IIII~~ III	IIII	II	~~IIII~~ I

How many marbles were in the bag? ______

Colour the spinner that is divided into 10 equal parts to represent the data in the frequency chart.

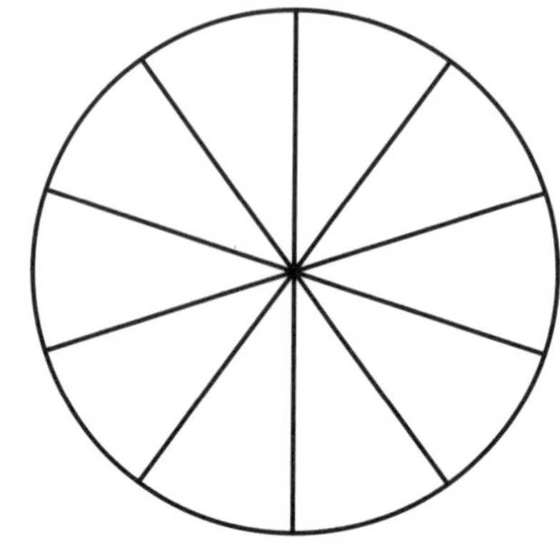

Red = $\frac{}{10}$
Blue = $\frac{}{10}$
Green = $\frac{}{10}$
Gold = $\frac{}{10}$

Number and Algebra

SET 3 Equivalent fractions

Colour the diagrams on the right side so that you have a pair of equivalent fractions.

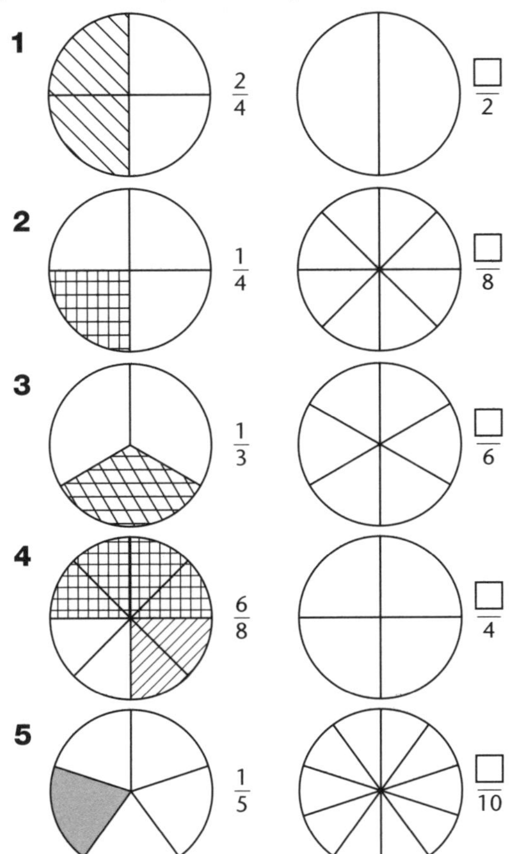

True or false?

6 $\frac{2}{3}$ is greater than $\frac{3}{6}$ ________

7 $\frac{3}{4}$ is greater than $\frac{5}{8}$ ________

8 $\frac{3}{5}$ is greater than $\frac{3}{10}$ ________

SET 4 Extension

1 1.25 km = ☐ m

2 Share $357 among 6.

3 7 × 501

4 How many eggs in $9\frac{1}{2}$ dozen?

5 How many faces does a triangular prism have?

6 What is the perimeter of a hexagon with 16 cm sides?

7 Centimetres in 13.75 m

8 Which one is not equivalent: $\frac{1}{4}$, 25%, 0.35 or $\frac{25}{100}$?

Working Mathematically

Round these numbers to the nearest 100 to estimate an answer to the multiplications.

	Number	Rounded to	Multiplied by	Estimate
9	321	300	4	1200
10	494		8	
11	340		6	
12	897		3	
13	515		5	
14	899		9	

Measurement Millimetres and centimetres

Measure the length of each line to the nearest millimetre.

1 __________________________________ ☐ mm

2 __________________________ ☐ mm

3 _____________________________________ ☐ mm

4 ___________________________________ ☐ mm

5 ____________________________________ ☐ mm

Write each centimetre measurement in millimetres.

6 3 cm ☐ **7** 7 cm ☐ **8** 18 cm ☐ **9** $9\frac{1}{2}$ cm ☐ **10** $11\frac{1}{2}$ cm ☐

UNIT 8

Number and Algebra

SET 1 Basic

1 Divide 40 by 5.

2 Product of 7 and 5

3 17 + 7

4 3 × 9

5 19 ☐ 8 = 27

6 27 ☐ 3 = 9

7 38 – 8

8 21 ÷ 7

9 9 ☐ 3 = 27

10 4 m = ☐ cm

11 Which is greater, $\frac{7}{10}$ or $\frac{7}{100}$?

12 Factors of 15

13 Are 35 and 50 multiples of 5?

14 How many 10c lollies can I buy for $1.50?

15 

Maria drank four 250 mL glasses of juice. How many litres is this?

☐ L

SET 2 Addition of 3-, 4- and 5-digit numbers

Populations

SHELLHARBOUR 52 800	GERRINGONG 2891
MINNAMURRA 534	NOWRA 23 823
KIAMA 5206	ULLADULLA 10 698

Find the combined populations of these towns.

1

SHELLHARBOUR	
MINNAMURRA	

2

ULLADULLA	
NOWRA	

3

NOWRA	
KIAMA	

4

GERRINGONG	
SHELLHARBOUR	

5 What is the total population of the six towns?

Measurement Perimeter

Measure the edges of the shapes in millimetres. Record the perimeter in the boxes.

1 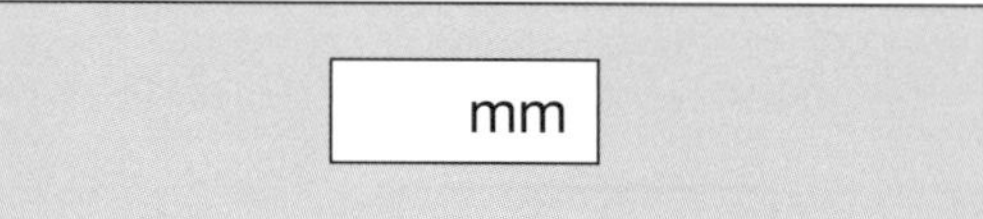

☐ mm

2 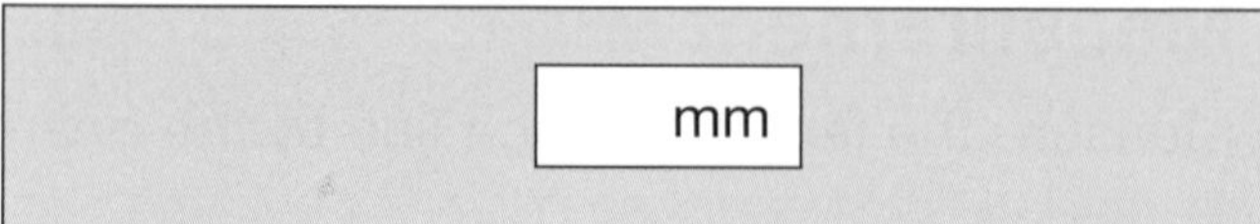

☐ mm

3

☐ mm

4 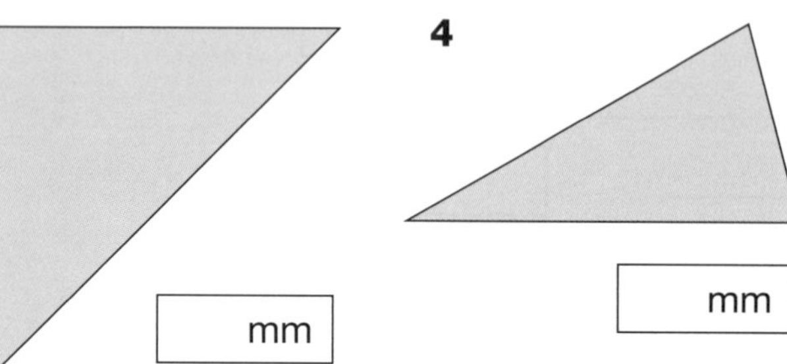

☐ mm

5 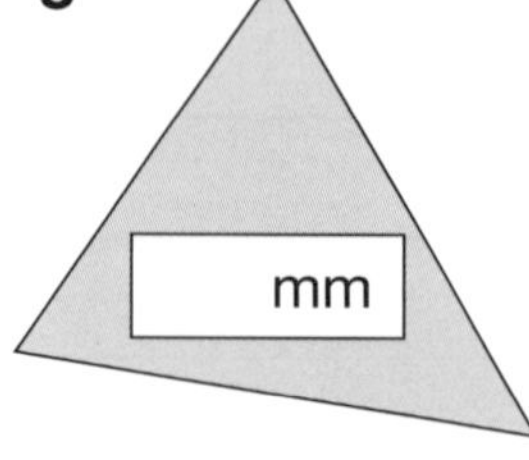

☐ mm

6 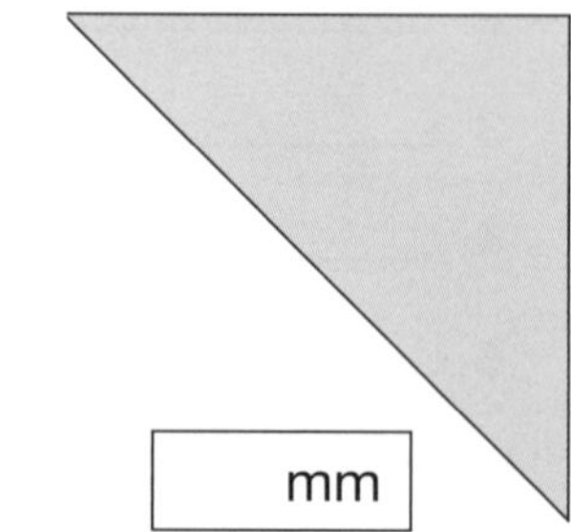

☐ mm

Number and Algebra

SET 3 Balance

Complete the number sentences by supplying the missing numbers.

1 6 + ☐ = 22 – 7

2 8 × ☐ = 9.6

3 36 ÷ ☐ = 54 ÷ 9

4 9 × ☐ = 13.5

5 ☐ × 4 = 75 – 55

6 13 × ☐ = 49 – 10

7 17 + ☐ = 96 – 66

8 7 × ☐ = 3.5

9 ☐ ÷ 7 = 81 – 75

10 6.4 ÷ ☐ = 0.8

Working Mathematically

Supply the missing operation signs.

11 8 ☐ 5 ☐ 2 = 42

12 (8 ☐ 5) ☐ 2 = 26

13 (8 ☐ 5) ☐ 2 = 6

14 8 ☐ (5 ☐ 2) = 24

SET 4 Extension

1 Write 13.25 as a mixed number.

2 Average of 46, 23, 69 and 22

3 How many kg in 3.1 tonnes?

4 Which is larger, $\frac{3}{4}$ or $\frac{3}{10}$?

5 Which temperature would be a cool winter's day: 41°C, 30°C, 100°C or 15°C?

6 Write one million, two hundred and twenty-six thousand in figures.

7 How many vertices on 3 cubes?

8 How many faces has a triangular pyramid?

9 (Triangle with angles x, 60°, 60°) x = ☐ °

10 Order 3.5, $4\frac{1}{4}$, 3.75 and $3\frac{9}{10}$.

11 How much is 3.2 kg of meat at $10 per kg?

12 How many axes of symmetry has a hexagon?

Working Mathematically

13 How many children are in our class if $\frac{2}{3}$ of the class are girls and half of them make a squad of 11 netball players?

____

Statistics and Probability Dot plots

www.Cars4You.Kombi created a dot plot to record their vehicle sales over 6 months.

1 How many sedans were sold?

2 How many 4WDs were sold?

3 How many trucks were sold?

4 How many more 4WDs were sold compared to utes?

5 How many more sedans were sold compared to trucks?

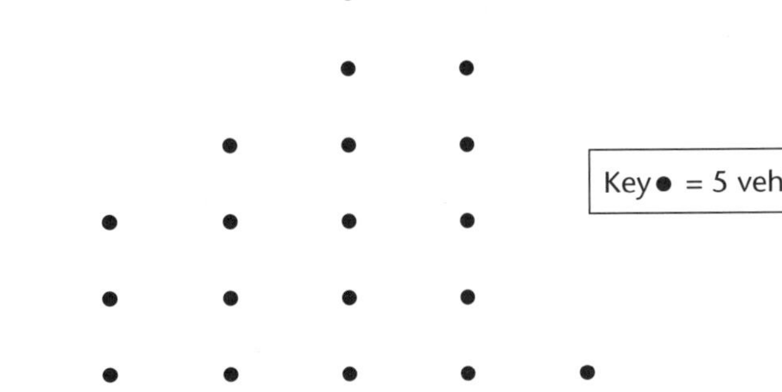

UNIT 9

Number and Algebra

SET 1 Basic

1 7^2

2 3 ☐ 8 = 24

3 35 ☐ 5 = 7

4 8 × 8

5 36 – 8

6 37 + 8

7 42 ÷ 6

8 Product of 9 and 7

9 Sum of 15 and 23

10 Cents in $37.52

11 Are 21 and 16 multiples of 6?

12 $2\frac{3}{4}$ hours = ☐ minutes

13 Which is greater, 27 hundredths or 30%?

14 55c × 6

15

SET 2 Division with fractional remainders

1 $3\overline{)6129}$

2 $5\overline{)1845}$

3 $7\overline{)7861}$

4 $4\overline{)2632}$

5 $6\overline{)2322}$

6 $10\overline{)6510}$

Record the remainders as fractions.

7 $5\overline{)6883}$

8 $4\overline{)7302}$

9 $3\overline{)5857}$

10 $6\overline{)7394}$

11 $8\overline{)6603}$

12 $5\overline{)7264}$

13 A box of 1457 lollies was sorted into 4 piles. How many in each pile?

14 728 Christmas lights were put into 5 packets. How many lights were in each packet?

Space Grid references

Plot, then join the grid references to discover the object.

(C,6) (G,5)
(G,5) (G,3)
(G,3) (H,1)
(H,1) (I,3)
(I,3) (I,5)
(I,5) (M,6)
(M,6) (I,6)
(I,6) (I,8)
(I,8) (J,9)
(J,9) (F,9)
(F,9) (G,8)
(G,8) (G,6)
(G,6) (C,6)

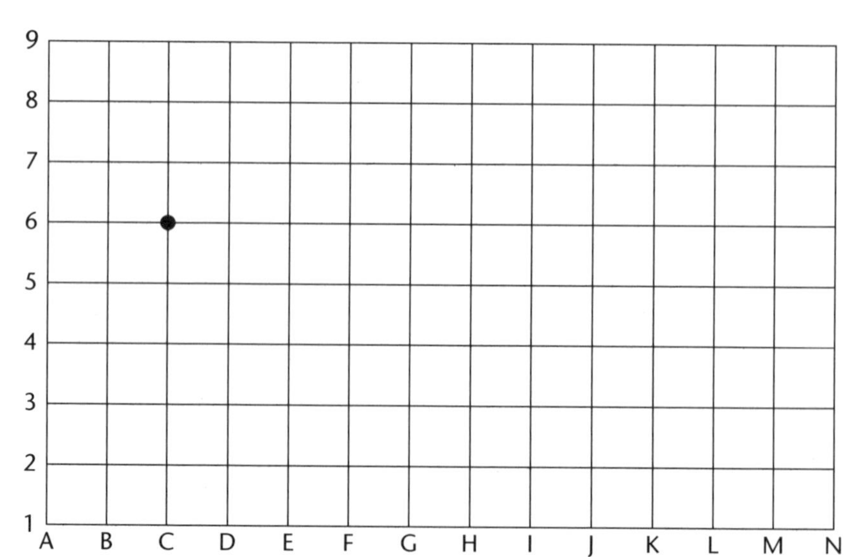

Number and Algebra

SET 3 Improper fractions and mixed numerals

True or false?

1 $\frac{3}{2} = 1\frac{1}{2}$ ☐ **2** $\frac{5}{4} = 1\frac{1}{4}$ ☐

3 $\frac{15}{10} = 1\frac{5}{10}$ ☐ **4** $\frac{6}{5} = 1\frac{1}{5}$ ☐

5 $\frac{4}{3} = 1\frac{2}{3}$ ☐ **6** $\frac{9}{4} = 2\frac{1}{4}$ ☐

7 $\frac{12}{5} = 2\frac{2}{5}$ ☐ **8** $\frac{8}{5} = 1\frac{4}{5}$ ☐

9 $\frac{7}{3} = 2\frac{1}{3}$ ☐ **10** $\frac{7}{2} = 3\frac{1}{2}$ ☐

Convert each improper fraction to a mixed numeral.

11 $\frac{4}{3} =$ **12** $\frac{7}{5} =$

13 $\frac{9}{4} =$ **14** $\frac{8}{6} =$

15 $\frac{11}{5} =$ **16** $\frac{5}{2} =$

Name two improper fractions that are equal to each mixed numeral.

	Mixed numeral	Improper fractions	
17	$2\frac{1}{2}$		
18	$1\frac{1}{4}$		
19	$1\frac{1}{3}$		
20	$2\frac{1}{6}$		

Working Mathematically

SET 4 Extension

1 Centimetres in 3.7 m

2 30 923, 30 950, 30 977, ☐

3 How many faces has a pentagonal prism?

4 Average of 124, 148, 207 and 21

5 Round 19 978 to the nearest 1000.

6 $1\frac{3}{4}$ min = ☐ seconds

7 33 + 27 ÷ 3 – 9 =

8 Which one is not equivalent: 0.5, 50%, $\frac{1}{2}$ or 0.51?

9 2.35 km = ☐ m

10 Write two hundred and ninety thousand and six in figures.

11 If 6 erasers cost $2.40, how much would 5 cost?

12 How much is 4.25 m of ribbon at $8 per metre?

Statistics and Probability Chance

1 Four marbles, 2 red and 2 yellow, are put into a bag. Write 5 different ways that the marbles can be drawn out of the bag.

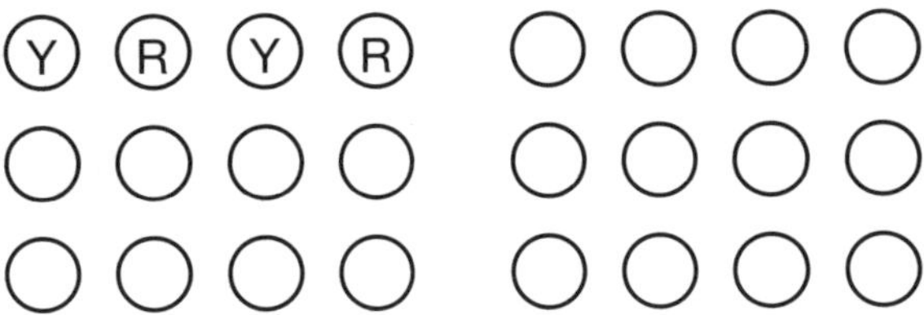

Two sets of dice have been rolled showing the total scores of 7 and 10. One example of each score is shown below.

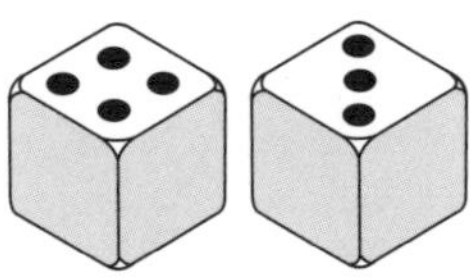

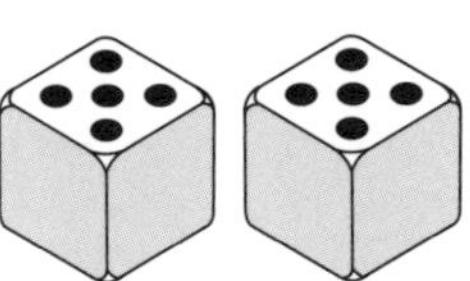

2 Which score do you think is more likely to occur? Explain why. ______________________

____

____

Number and Algebra

SET 1 Basic

1 7 + 9

2 8 × 7

3 24 – 9

4 49 ☐ 40 = 89

5 7 ☐ 10 = 70

6 5^2

7 Divide 35 by 5.

8 50 ☐ 25 = 25

9 Product of 7 and 6

10 Quotient of 45 and 5

11 Cents in $13.24

12 (8 + 1) × 3

13 Is 27 a prime number?

14 3 L and 175 mL = ☐ mL

15 How many 25 cm lengths can be cut from 1 m?

16

Dominic has 25 books to read in a Read-A-Thon. If he has read 13, how many are left to read?

☐ books

SET 2 Geometric patterns

Complete the matchstick number patterns.

1  

Squares	1	2	3	4	5	6	7	8
Matches	4							

2  

Triangles	1	2	3	4	5	6	7	8
Matches	3							

Triangular numbers can be displayed in the shape of a triangle.

3 Name the next 5 triangular numbers.

☐ ☐ ☐ ☐ ☐

Space Triangles

Name each triangle.

1

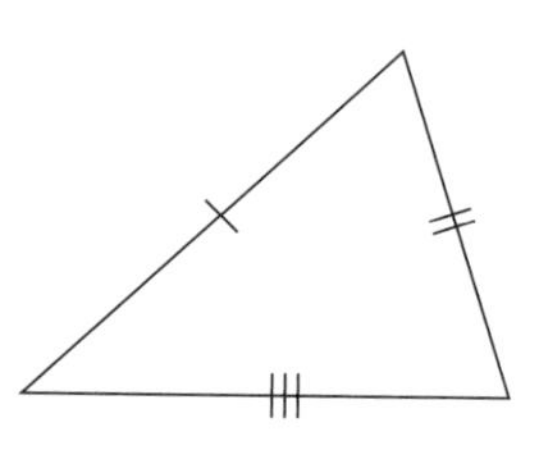

____

2

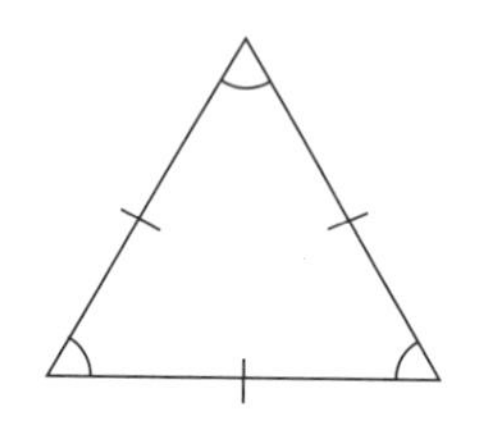

____

3

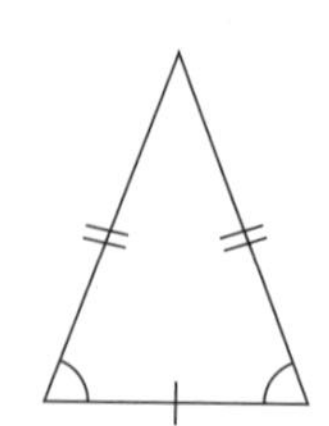

____

4

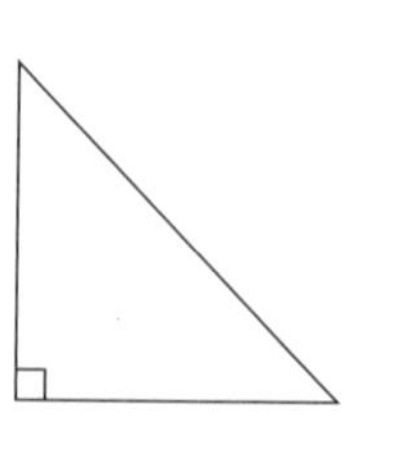

____

Number and Algebra

SET 3 Prime and composite numbers

Write prime or composite after each number.

	Number	Prime or composite
1	11	
2	20	
3	21	
4	44	
5	47	
6	51	
7	99	
8	53	
9	41	
10	92	

State the prime numbers that are found between each pair of numbers.

11 6 and 10 ________

12 15 and 20 ________

13 20 and 30 ________

14 50 and 60 ________

15 90 and 110 ________

SET 4 Extension

1 Average 10, 90, 170, 90

2 $\frac{3}{10} = \frac{30}{100}$. True or false?

3 4.75 m = ☐ cm

4 350 – 90 ÷ 5

5 How many sides has a decagon?

6 3.52 + 7.64

7 How many faces has a hexagonal prism?

8 $\frac{3}{4}$ of 200

9 At what temperature would water boil: 0°C, 25°C, 50°C or 100°C?

10 Write one hundred and five thousand, two hundred and twenty-six in figures.

11 37 280, 37 250, 37 220, ☐

12 How many axes of symmetry has an isosceles triangle?

13 Which one is not equivalent: $\frac{3}{4}$, $\frac{75}{100}$, 80% or 0.75?

14 Find the mean population of the following towns:

Shellharbour	52 800	Gerringong	2891
Minnamurra	534	Nowra	23 823
Kiama	5206	Ulladulla	10 698

Working Mathematically

Measurement Area of triangles

Find the area of the rectangle, then halve it to find the area of the triangle.

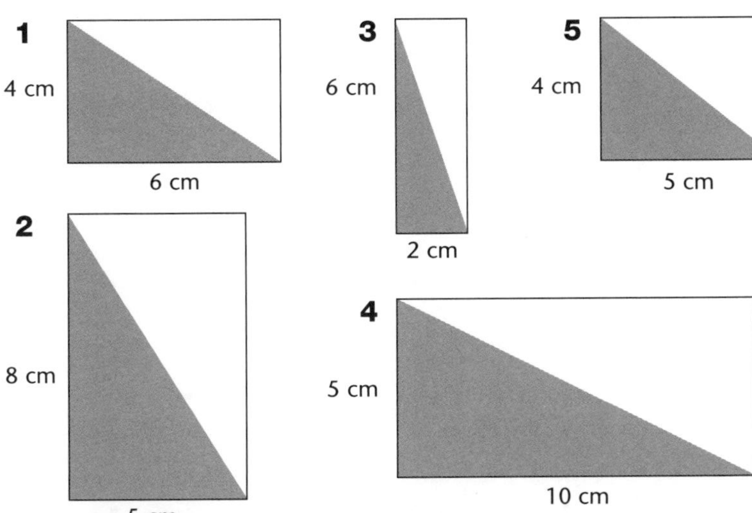

	Area of ▭	Area of ◺
1		
2		
3		
4		
5		

UNIT 11

Number and Algebra

SET 1 Basic

1 $25 \div 3$

2 $15 \div 3 + 10$

3 Which is the 23rd letter of the alphabet?

4 $6^2 + 3$

5 Days in 2 years (not leap years)

6 $42 + 19$

7 Perimeter of a square 5 cm wide

8 $9 + 500 + 40$

9 $42 - 17$

10 $21\% = 0.$ ☐

11 $3 \times 9 \times 0$

12 $29 \div 9$

13 Difference between 9^2 and 20

14 $8 + 600 + 70$

15 One-third of 27

16

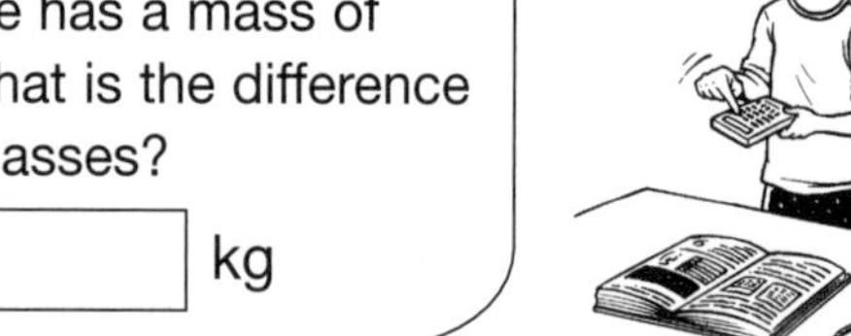

SET 2 Unit fractions of a quantity

Find these fractions of 36 stars.

1 $\frac{1}{2} =$

2 $\frac{1}{6} =$

3 $\frac{1}{4} =$

4 $\frac{1}{12} =$

5 $\frac{1}{9} =$

6 $\frac{1}{3} =$

Find the fraction of each collection.

7 $\frac{1}{3}$ of 90 =

8 $\frac{1}{5}$ of 100 =

9 $\frac{1}{4}$ of 120 =

10 $\frac{1}{5}$ of 200 =

11 $\frac{1}{5}$ of 300 =

12 $\frac{1}{4}$ of 240 =

13 $\frac{1}{8}$ of 240 =

14 $\frac{1}{5}$ of 400 =

15 $\frac{1}{8}$ of 400 =

16 $\frac{1}{5}$ of 800 =

17 $\frac{1}{8}$ of 640 =

18 $\frac{1}{10}$ of 600 =

Space Representing three-dimensional objects

Draw these prisms on the dot paper. The back block has been drawn for you.

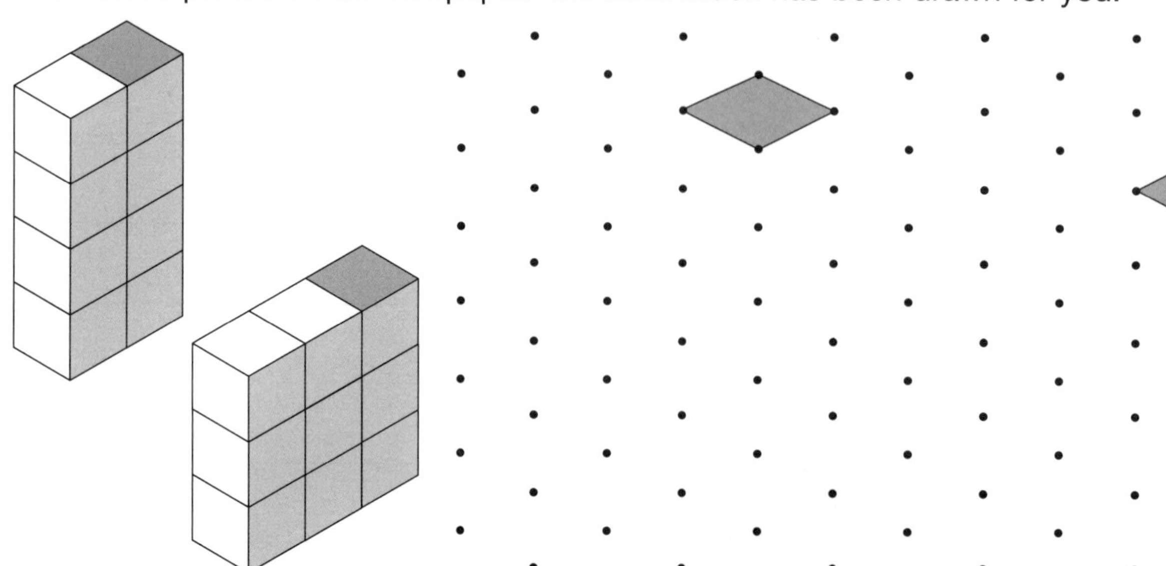

Number and Algebra

SET 3 Decimals to thousandths

State the place value of each bold digit. The words in the cloud may help you.

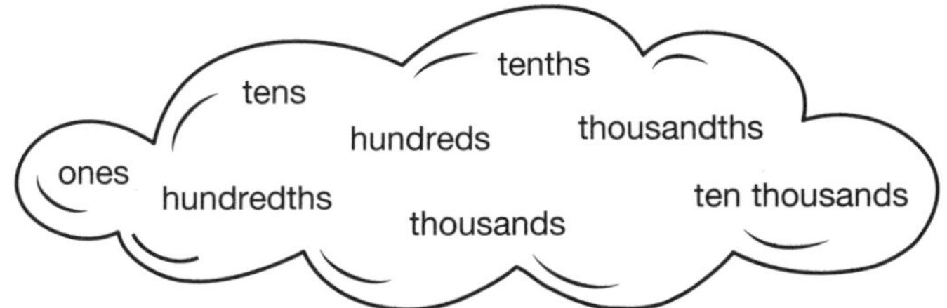

	Number	Place value
1	36**7**.361	
2	3**6**5.973	
3	**7**93.548	
4	357.**2**42	
5	679.3**8**4	
6	**3**574.261	
7	4729.3**7**5	
8	293.74**2**	

Write these mixed numerals as decimals.

	Mixed numeral	Decimal		Mixed numeral	Decimal
9	$7\frac{7}{10}$		11	$6\frac{291}{1000}$	
10	$3\frac{22}{100}$		12	$7\frac{7}{100}$	

SET 4 Extension

1 Which does not fit: $\frac{1}{2}$, 0.5, $\frac{22}{40}$ or 50%?

2 Write 5:10 pm in 24-hour time.

3 What is the volume of a box with length 5 cm, width 2 cm and height 6 cm?

4 How many minutes in 2.25 hours?

5 At what temperature does ice melt?

6 Area of a rectangle with sides 40 m and 8 m

7 How much are 12 cakes at 3 for $4.75?

8 How many faces has a square pyramid?

9

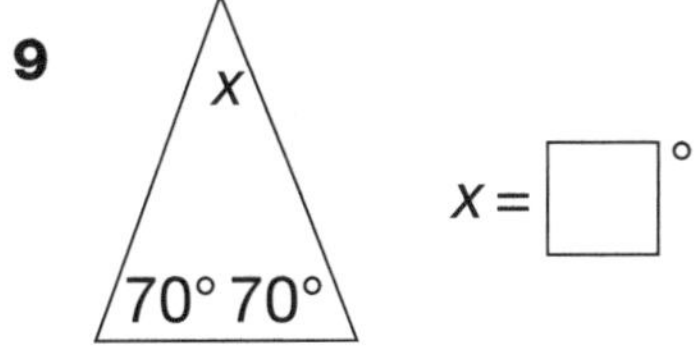

$x =$ ☐°

10 $40 less $5.50

11 4.35 km = ☐ m

12 Round 637 428 to the nearest 10.

13 How many axes of symmetry has a regular octagon?

14

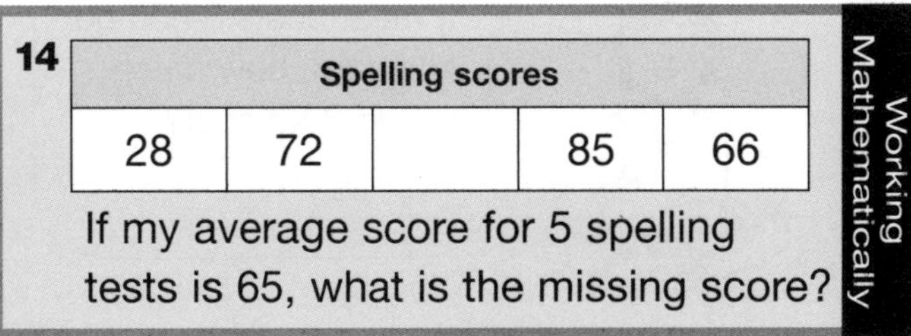

Spelling scores				
28	72		85	66

If my average score for 5 spelling tests is 65, what is the missing score?

Working Mathematically

Statistics and Probability Chance from 0 to 1

1 Describe the chance of pulling each coin out of the bag by drawing a line from the coin to the chance scale.

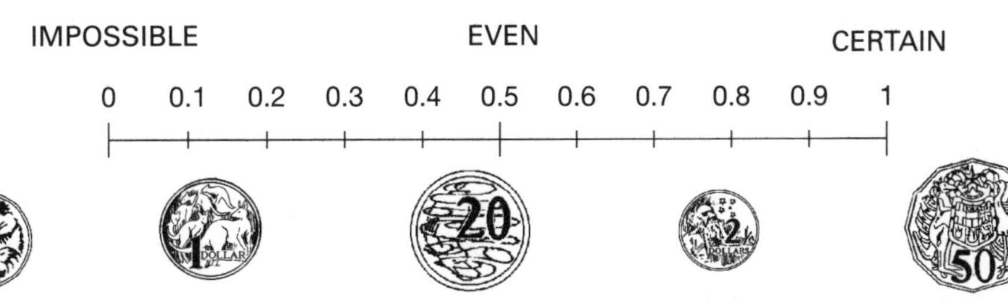

2 Describe the chance of pulling 2 coins out of the bag that have a combined value of less than 50c.

UNIT 12

Number and Algebra

SET 1 Basic

1 150 + 30

2 20 × 6

3 600 ☐ 550 = 50

4 27 ☐ 12 = 15

5 3 ☐ 12 = 36

6 43 ÷ 6

7 107 ☐ 107 = 214

8 Product of 7 and 0

9 Sum of 127 and 20

10 0.9 as a fraction

11 Value of 6 in 364 000

12 Is 15 a prime number?

13 4 mins = 240 sec. True or false?

14 3 L and 270 mL = ☐ mL

15 $5^2 + 6$

16 

If meat costs $4.50 per kilogram, how much would 3 kg cost?

$ ☐

SET 2 Adding decimals

1 3.2 + 2.6

2 8.7 – 3.5

3 3.2 + 7.4 + 14.3

4 1 whole + 3 tenths + 2 hundredths

5 Order these decimals: 13.5, 13.6, 13.71, 13.65

6 18.6 – 14.4

7 Which is larger, 1 whole or 1.1?

8 265.84 + 1.03

9 $48.56 + $21.13

10 $312.50 + 432.50

11 $140.70 + 1.86

12 $364.24 + 128.36

13 $3642.20 + 168.27

14 $513.62 + 39.06 + 290.80

15 $768.37 + 6.93 + 405.08

Statistics and Probability Line graphs

1 How much did Kelly weigh at age 5?

2 How much did she weigh at age 13?

3 Between which ages did she not put on weight?

4 How much weight did she put on between the ages of 5 and 11?

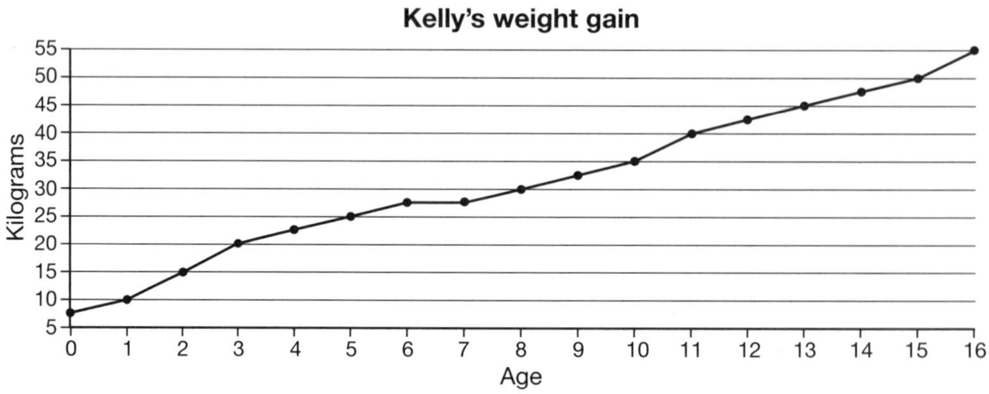

5 Estimate Kelly's weight at $10\frac{1}{2}$ years.

6 How much weight did she put on between the ages of 10 and 16?

Number and Algebra

SET 3 Decimals, percentages and fractions

Colour the strips to match the percentage.

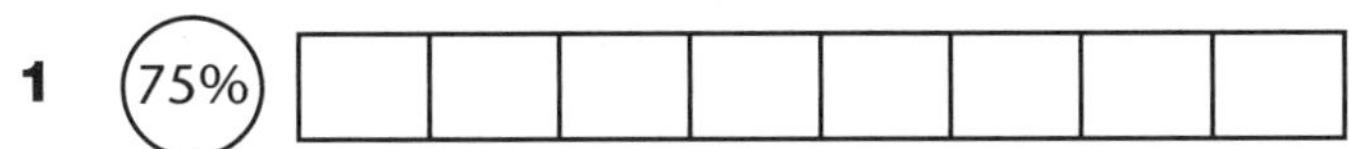

1 (75%)

2 (10%)

3 (25%)

4 (50%)

Express these percentages as decimals.

5 58% = ______ 8 72% = ______

6 45% = ______ 9 69% = ______

7 5% = ______ 10 99% = ______

Express these decimals as fractions.

11 0.5 = ______ 14 0.75 = ______

12 0.05 = ______ 15 0.25 = ______

13 0.1 = ______ 16 0.09 = ______

True or false?

17 25% > $\frac{3}{4}$ ______ 18 0.7 = 70% ______

SET 4 Extension

1 37% = 0.☐

2 Order 0.31, 30%, $\frac{9}{10}$, and $\frac{1}{5}$.

3 $\frac{8}{10} + \frac{8}{10} + \frac{7}{10}$

4 How much is 3.25 m at $20 per metre?

5 How many sides has a heptagon?

6 $\frac{25}{40} = \frac{\square}{8}$

7 3.5, 4, 4.5, 5, ☐

8 How many faces has an octagonal prism?

9 What is the perimeter of a square with sides of 3.5 cm?

10 If 20 pencils cost $2.60, how much would 30 cost?

11 2 dozen lollies at 6 lollies for 75c

12 How much is 4.2 m of timber at $5 a metre?

13 How many 3 cm cubes will fit into a box measuring 6 cm in length, 3 cm in width and 6 cm in height?

Working Mathematically

14 
How tall are Sally and Anna if Sally is 0.1 m taller than Anna and their combined height is 3.16 metres?

Sally: ______ Anna: ______

Measurement Calculating volume

What is the volume of each box?

Volume = length × width × height

1

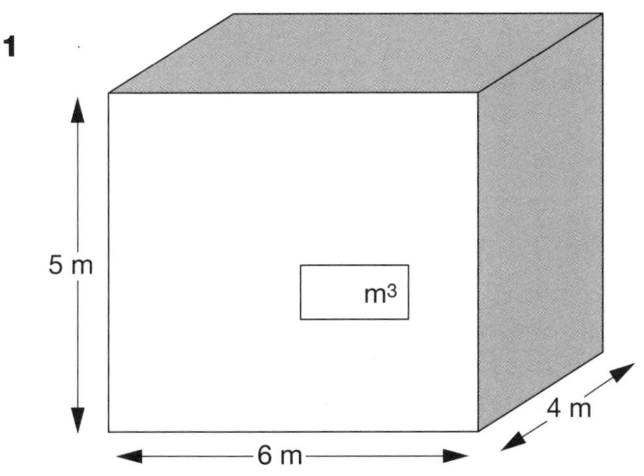

2

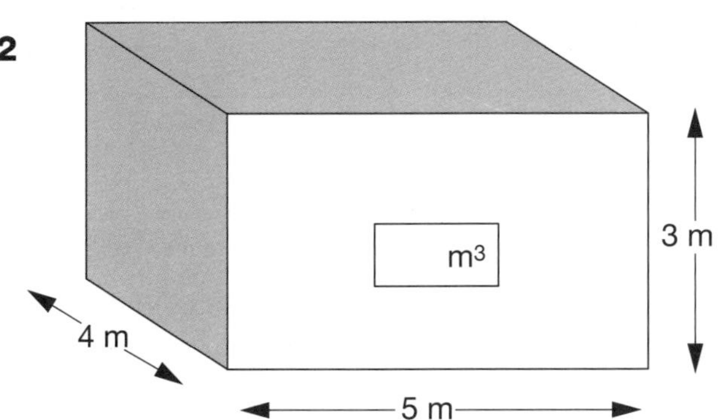

UNIT 13

Number and Algebra

SET 1 Basic

1 18 ☐ 12 = 30

2 100 ☐ 90 = 10

3 6^2

4 Divide 18 by 2.

5 7 × 7

6 24 – 8

7 18 + 21

8 26 + 17

9 Are 14 and 31 multiples of 6?

10 Is 7 a prime number?

11 Divide 13 by 6.

12 Value of 3 in 35 407

13 45c × 3

14 3 × 7 + 9

15 How many 500 g bags make 2 kg?

16

Cook House scored 650 points and received 45 bonus points. How many points did Cook have?

☐ points

SET 2 Operations with decimals

1 3.27 + 2.65

2 8.76 – 3.54

3 13.45 + 4.54 + 6.55

4 \$14.45 × 2

5 \$15 – \$12.45

6 (\$5.65 + \$3.35) ÷ 3

7 5.787 km + 3.202 km

8 \$100 – (\$25.50 + \$20.10)

9 Share \$48.56 among 4

10 Cost of 6 tickets at \$3.20 each

11 \$15.65 + ______ = \$20.20

12 7.536 metres ÷ 6

Mathematical Reasoning

Colonel Saunders recorded the price of hamburgers in 5 countries.

USA	England	Australia	China	Russia
\$3.00	\$3.40	\$3.20	\$1.45	\$2.45

Work out the difference in price if you were buying hamburgers in these countries.

13	USA and China	
14	England and Russia	
15	Australia and China	
16	England and China	

Space Drawing rectangles and squares

Use the 5 mm dot paper to draw these rectangles.

1 

2 cm

4 cm

2 

3 cm

5 cm

3 

3 cm

3 cm

Number and Algebra

SET 3 Equivalent number sentences

Find the missing numbers to complete the number sentences.

1 $5 \times \square = 45$

2 $\square + 7 = 60 - 8$

3 $23 + \square = 10 \times 5$

4 $17 + \square = 100 \div 4$

5 $\square \times 5 = 350 \div 10$

6 $36 \div \square = 62 - 56$

7 $81 \div \square = 54 \div 6$

8 $\square + 21 = 100 - 57$

9 $\square \times 6 = 200 - 164$

10 $9 \times 7 = \square + 50$

11 $(\square + 7) \times 9 = 135$

12 $3 \times \square + 6 \times 9 = 81$

Given that $\blacksquare = 8$, calculate the following.

13 $\blacksquare \times 7 + 8 - 16 =$

14 $(\blacksquare + 7) \times 3 \div 5 =$

15 $3 \times (\blacksquare + 12) - 37 =$

SET 4 Extension

1 Perimeter of a hexagon with 3.25 m sides

2 How much are 7 pencils at 3 for $1.65?

3 How much is 7.25 kg at $20 per kilogram?

4 Order 1.53, 150%, 1.49 and $1\frac{4}{10}$.

5 Add all the odd numbers between 20 and 30.

6 Is a hectare roughly equal to 2 tennis courts or to 2 soccer fields?

7 $\frac{7}{10}$ of 2.5 m = $\square$ cm

8 What is the area of a rectangle with sides 7 cm and 3.5 cm?

9 What fraction of 9 m is 150 cm?

10 How many cm^2 in 1 m^2?

11 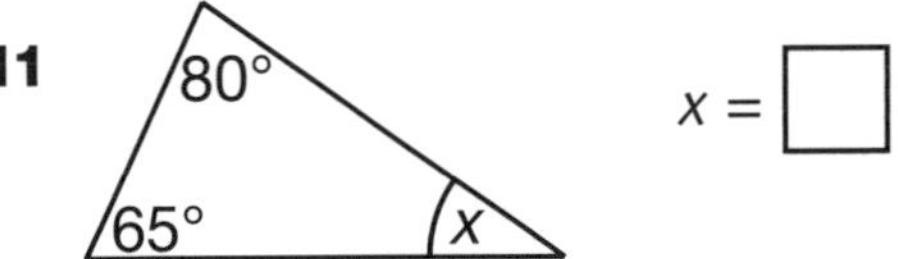

$x = \square°$

12 Average of $3.50, $4.50, $5.50 and $2.90

13 How many axes of symmetry has a rhombus?

14 How many mL in 5.75 L?

15 What object does this net make?

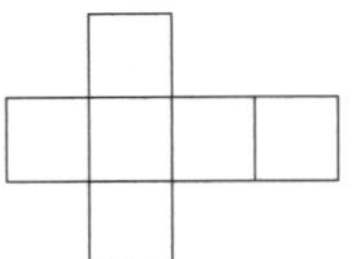

Measurement 24-hour time

Display each 24-hour time on a clock face and express each time in digital form underneath.

1

2118

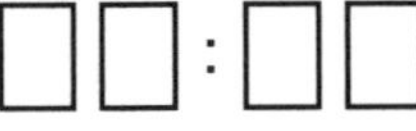

2 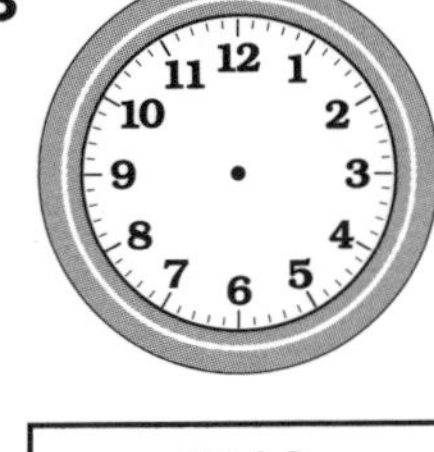

0405

$\square\square : \square\square$

3

1713

$\square\square : \square\square$

4 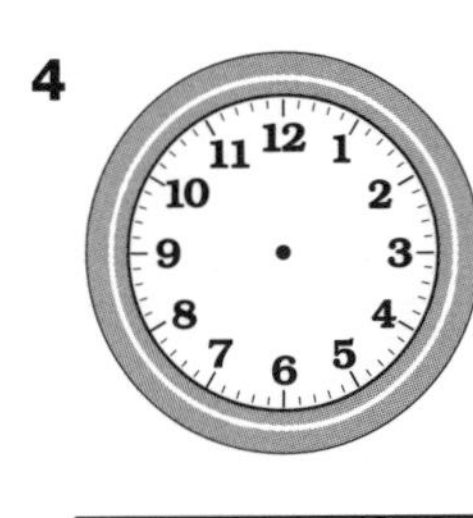

1344

5 

1142

$\square\square : \square\square$

Number and Algebra

SET 1 Basic

1 75 ☐ 50 = 25

2 $4^2 + 5$

3 23 – 8

4 9 ☐ 12 = 21

5 Divide 42 by 7.

6 $1\frac{1}{4}$ hours = ☐ minutes

7 One-third of 45

8 130 ÷ 10

9 Difference between $4 and $1.35

10 Value of 4 in 7043

11 Are 28 and 56 multiples of 7?

12 How many sides has an octagon?

13 How many 50c coins in $7?

14 13 km + 38 km

15 Sam saved $6 a week for 8 weeks. How much did he save? $ ☐

SET 2 Subtraction of 4- and 5-digit numbers

1 The distance between Albury and Tweed Heads is 1458 km. How far have I still to go if I've travelled 969 km? 

2 What is the difference in mass between HMAS *Hobart* (7000 t) and HMAS *Stuart* (3600 t)? 

3 Cassie saved $9410 to buy a car. The car she bought cost $8999. How much has she left?

4 How much is left in the tank if 31 150 L have been used out of the 50 000 L available?

5 How many tickets are left out of 24 500 if 2670 have already been sold?

Space Drawing objects from views

Draw the top, front and side views of each object.

	Top view	Front view	Side view
1			
2			

Number and Algebra

SET 3 Comparing and ordering fractions

1 Draw a line to your estimate of the place of each fraction on the number line.

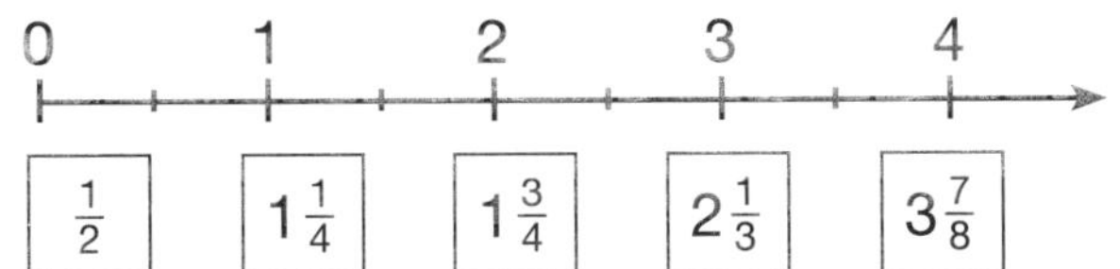

2 Label each place on the number line.

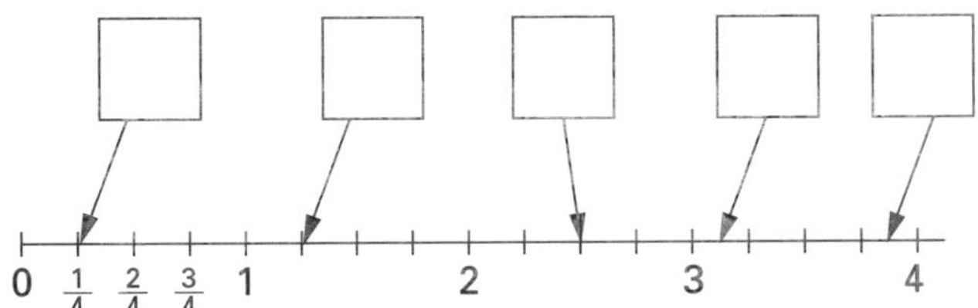

Continue the sequences.

3	$3\frac{1}{2}$	$4\frac{1}{2}$	$5\frac{1}{2}$			
4	$7\frac{1}{4}$	$7\frac{1}{2}$	$7\frac{3}{4}$			
5	$8\frac{1}{3}$	8	$7\frac{2}{3}$			
6	$6\frac{1}{4}$	$6\frac{3}{4}$	$7\frac{1}{4}$			
7	7	$7\frac{3}{4}$	$8\frac{1}{2}$			
8	$5\frac{3}{5}$	$6\frac{1}{5}$	$6\frac{4}{5}$			
9	$7\frac{7}{10}$	$8\frac{3}{10}$	$8\frac{9}{10}$			

SET 4 Extension

1 $\frac{3}{4}$ of $240

2 How many faces has a hexagonal pyramid?

3 3 dozen cakes at 45c each

4 If 16 bananas cost $2.40, how much would 20 cost?

5 Estimate 396 × 31.

6 Write 1:15 pm in 24-hour time.

7 7.4 m = ☐ cm

8 $\frac{20}{45} = \frac{4}{\square}$

9 What fraction of 300 is 25?

10 How many 1.4 m lengths can be cut from a 7 m piece?

11 How many 300 g bags in $4\frac{1}{2}$ kg?

12 How many 6 cm squares are needed to cover a rectangle with sides 12 cm and 24 cm?

13 How much is 2.25 kg at $4.40 per kilogram?

14 The park and the pre-school each occupy one hectare of land, but are different shapes. Which one has the larger perimeter if the pre-school is a square with 100-metre sides and the park is a rectangle 200 metres long and 50 metres wide?

Working Mathematically

Measurement Metres, centimetres and millimetres

Estimate these measurements.

1	The length of your classroom	
2	The length of your bedroom	
3	The length of your pencil	
4	The length of a sultana	
5	The length of your shoe	
6	The distance from school to home	
7	The height of your teacher	
8	The width of a wedding ring	

Number and Algebra

SET 1 Basic

1 One position after 21st

2 $9 + 2\frac{1}{2}$

3 $16 \div 5$

4 18 ☐ 3 = 6

5 40 mm + 5 cm = ☐ mm

6 Next ordinal number after 3rd

7 What is the date one week after Anzac Day?

8 Difference between 16 and 3^2

9 $\frac{1}{2}$ of \$0.70

10 \$0.24 × 2

11 $200 \div 10$

12 Square 5.

13 Cost of ten 65c stamps

14 8^2

15 

How much are 10 books at \$1.25 each?

\$ ☐

SET 2 4-digit multiplication

1 1342 × 4

2 2245 × 3

3 4173 × 5

4 2341 × 6

5 1054 × 8

6 5183 × 7

7 3718 × 4

8 6049 × 5

9 2244 × 9

10 A company bought 8 laptops for their executive staff. If each laptop cost \$5329, what was the total cost? 

\$ ☐

11 7950 books were sold at Christmas and 1262 at Easter. What was the total sales value if the books averaged \$9 each? 

\$ ☐

Statistics and Probability Potentially misleading data

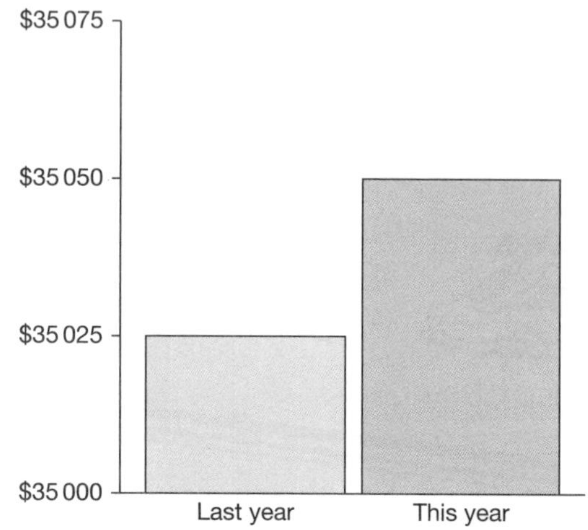

Explain how the data presented in the graph and the graph's title could be misleading.

UNIT 15

Number and Algebra

SET 3 Add and subtract fractions

Add these fractions.

1 $\frac{3}{8} + \frac{2}{8} =$

2 $\frac{7}{10} + \frac{1}{10} =$

3 $\frac{3}{10} + \frac{3}{10} + \frac{3}{10} =$

4 $\frac{3}{8} + \frac{3}{8} + \frac{1}{8} =$

5 $\frac{7}{8} + \frac{5}{8} = \quad =$

6 $\frac{3}{4} + \frac{3}{4} = \quad =$

7 $\frac{7}{10} + \frac{7}{10} = \quad =$

8 $\frac{3}{5} + \frac{4}{5} = \quad =$

9 $\frac{7}{10} + \frac{7}{10} + \frac{7}{10} = \quad =$

10 $\frac{3}{5} + \frac{4}{5} + \frac{2}{5} = \quad =$

Subtract these fractions.

11 $\frac{7}{10} - \frac{1}{10} =$

12 $\frac{7}{8} - \frac{2}{8} =$

13 $\frac{9}{10} - \frac{5}{10} =$

14 $\frac{4}{5} - \frac{3}{5} =$

15 $\frac{3}{4} - \frac{2}{4} =$

16 $\frac{5}{10} - \frac{2}{10} =$

17 $\frac{4}{8} - \frac{3}{8} =$

18 $\frac{6}{10} - \frac{1}{10} =$

19 $\frac{2}{5} - \frac{1}{5} =$

20 $\frac{9}{10} - \frac{3}{10} =$

SET 4 Extension

1 $19.15 = ☐ c

2 Share $16.50 among 3.

3 If tomatoes are $1.80 a kg, how much would 4500 g cost?

4 $36 + 17 + 9 \times 7$

5 $46 + 7 \times 8 + 3$

6 If Mum drove 260 km in three hours, was her average speed about 90 km/h or 26 km/h?

7 How many metres in 5.25 km?

8 Change from $10 if I bought 3 bracelets at a cost of $2.70 each

9 Arrange in ascending order: 5.0, 0.5, 5^2, 15.0.

10 7.3 kg at $20 per kilogram

11 How many degrees between south and north-west?

Working Mathematically

12 Yolande said, 'If you double the sides of a cube you double its volume.'

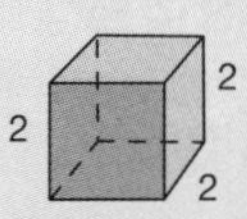

Test Yolande's theory and explain your results.

____

____

Measurement and Space Area of a parallelogram

Multiply the height of each parallelogram by the length of its base in order to find the area of each parallelogram.

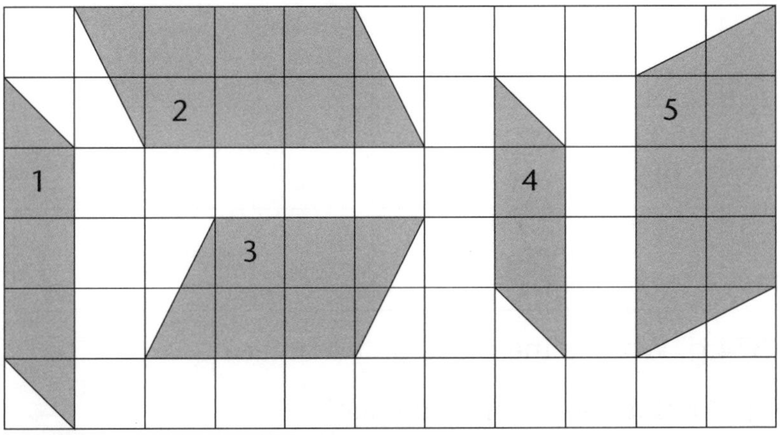

	Base × height	=	Area
1	×	=	cm^2
2	×	=	cm^2
3	×	=	cm^2
4	×	=	cm^2
5	×	=	cm^2

Number and Algebra

SET 1 Basic

1 9^2

2 42 + 19

3 76 – 38

4 62c × 4

5 $2\frac{3}{4}$ hours = ☐ minutes

6 2 L and 25 mL = ☐ mL

7 Which is larger, $2.50 × 10 or $250?

8 Add $\frac{6}{10}$ to 3.37.

9 Cost of nine 60c stamps

10 $\frac{19}{100}$ = 0.☐

11 Difference between 7^2 and 19

12 80 + 1000 + 400 + 3

13 Seconds in 10 minutes

14 18 ÷ 9 + 10

15 One-fifth of 45

16 A bricklayer lays 2 bricks a minute. How many will he lay in 45 minutes? ☐ bricks

SET 2 Extended multiplication

1 25×16

2 31×15

3 28×45

4 76×48

5 80×26

6 75×33

7 382×27

8 251×72

9 462×54

10 474×43

11 586×29

12 648×62

Measurement Mass and capacity

1 Complete the chart to show the relationships between the volume, capacity and mass of water.

Volume	300 cm^3			1000 cm^3			2000 cm^3			
Capacity		600 mL		1 litre	1500 mL			2415 mL		3 litres
Mass			900 g	1 kilogram		1900 g			2820 g	

2 A 250 mL bottle of water has a mass of 374 g. What is the mass of the bottle?

3 A 1.5 L jug of water has a mass of 1635 g. What is the mass of the jug?

Number and Algebra

SET 3 Expanding numbers

Expand these numbers. The first one is done for you.

1 37 423 = 30 000 + 7000 + 400 + 20 + 3

2 85 616 =

3 25 209 =

4 106 015 =

5 6 041 500 =

What number am I?

6 I am 40 000 more than 28 499

7 I am 6 000 more than 148 347

8 I am 90 more than 265 268

9 I am 700 more than 70 658

10 I am 2 000 000 less than 56 812 405

Write the following numbers.

11 400 000 + 80 000 + 200 + 7 =

12 4 000 000 + 300 000 + 10 000 + 100 =

13 5 000 000 + 500 000 + 40 000 + 1000 + 40 =

SET 4 Extension

1 Round and estimate $2.95 × 18.

2 $\frac{4}{5}$ of $300

3 $\frac{4}{8} + \frac{1}{8} + \frac{7}{8}$

4 How much are 6 books at 5 for $7.50?

5 9.49 – 4.61

6 20% of $150

7 Perimeter of a decagon with 7.5 cm sides

8 Half of $5\frac{1}{2}$ dozen

9 60% of 90

10 How many fifths in 7.4?

11 Order $1\frac{3}{5}$, 1.61, 1.5 and $1\frac{4}{10}$.

12 $50 – $39.90

13 $(8^2 - 7^2) \times 2$

14 70% of 1 tonne

15 What fraction of 3 km is 200 m?

16 Volume of a box with length 7 cm, width 4 cm and height 8 cm

17 Complete the grid to convert these improper fractions to mixed numerals.

$\frac{3}{2}$	$\frac{9}{8}$	$\frac{7}{5}$	$\frac{6}{4}$	$\frac{5}{3}$	$\frac{15}{10}$	$\frac{5}{4}$	$\frac{7}{3}$
$1\frac{1}{2}$	$1\frac{1}{8}$						

Space Rotational symmetry

Colour the shapes that have rotational symmetry.

1

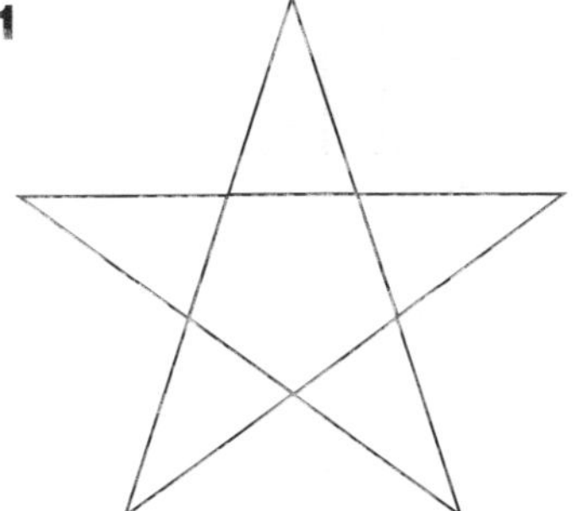

2

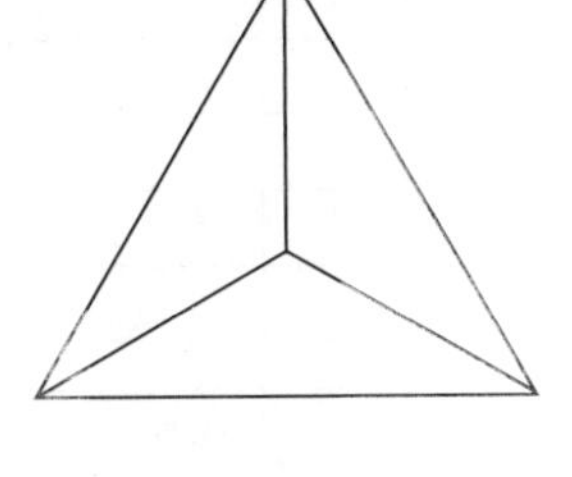

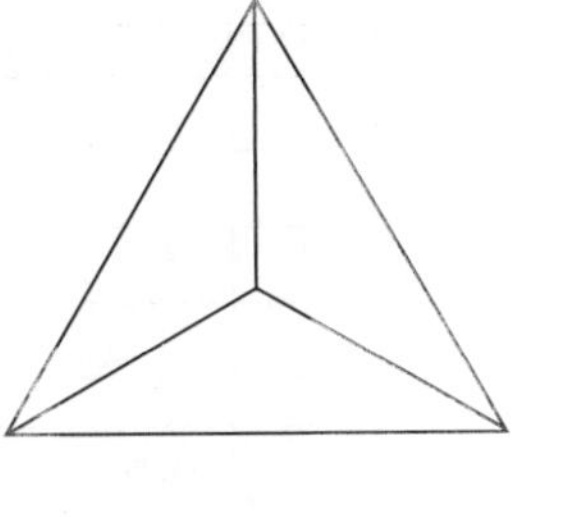

3

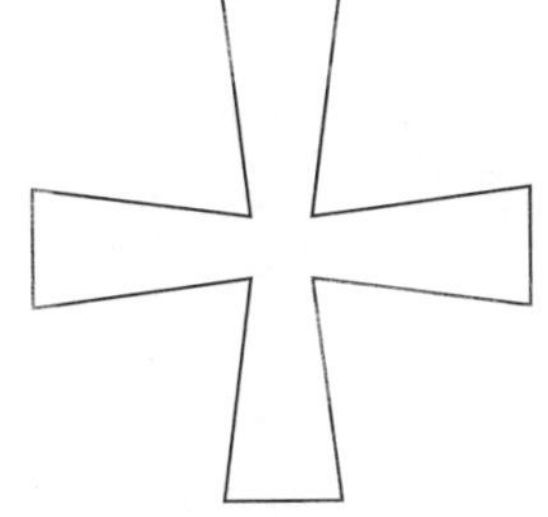

4

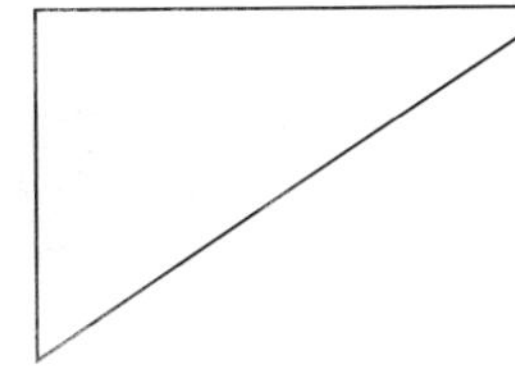

5

6

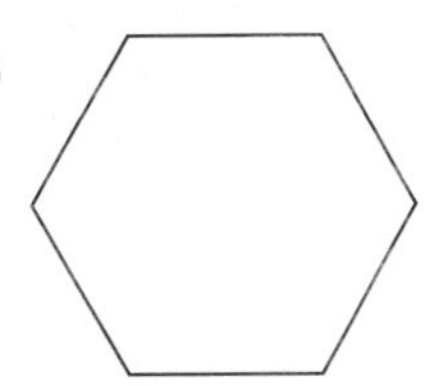

Number and Algebra

SET 1 Basic

1 11 + 31
2 7 × 9
3 35 – 6
4 28 ÷ 7
5 250 ☐ 2 = 500
6 500 ☐ 10 = 50
7 Product of 8 and 10
8 Divide 88 by 11.
9 Sum of 88 and 11
10 $5^2 + 7$
11 13, 26, 39, ☐
12 Is 43 a prime number?
13 Are 200 and 16 multiples of 6?
14 5 hundreds + 1387
15 How many 500 g packs are needed to fill 2 kg container?
16

SET 2 Order of operations

Use the order of operations to solve the questions.

1 7 × 3 + 20 =
2 7 × (3 + 20) =
3 7 × 5 + 9 – 6 =
4 35 – 3 × 5 + 6 =
5 5 × (9 + 6) + 30 =
6 35 + 7 – 6 × 3 + 40 =
7 66 ÷ 6 + 7 + 40 =
8 200 – 7 × 6 + 30 =
9 300 – 3 × 8 × 2 + 7 =
10 400 + 7 × 3 – 200 =
11 (90 – 45) × 5 – 170 =
12 3 × 5 × 2 × 4 × 4 ÷ 8 – 59 =
13 500 – 3 × 4 × 3 × 5 × 2 ÷ 2 =

14 Kiet saved $497 in May, $256 in June and $387 in July towards a new TV set. How much more does he need to save if the TV costs $1497? 

Space Top, front and side views

Draw the top, front and side views of each object.

1

Top	Front	Side

2

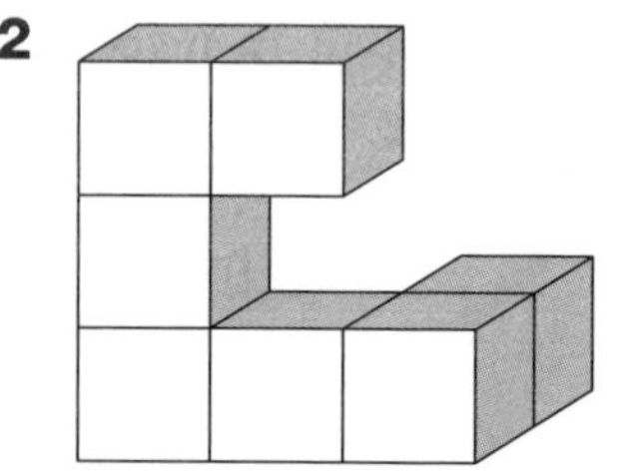

Top	Front	Side

Number and Algebra

SET 3 Add and subtract fractions

1 $\frac{3}{4} - \frac{1}{4} =$

2 $\frac{6}{10} - \frac{1}{10} =$

3 $\frac{9}{10} - \frac{3}{10} =$

4 $\frac{7}{8} - \frac{5}{8} =$

5 $\frac{7}{8} - \frac{4}{8} =$

Record these answers as improper fractions and as mixed numerals.

6 $\frac{8}{10} + \frac{5}{10} =$ $\quad =$

7 $\frac{4}{5} + \frac{3}{5} =$ $\quad =$

8 $\frac{7}{10} + \frac{4}{10} =$ $\quad =$

9 $\frac{8}{10} + \frac{9}{10} =$ $\quad =$

10 $\frac{2}{4} + \frac{3}{4} =$ $\quad =$

11 $\frac{3}{4} + \frac{3}{4} + \frac{1}{4} =$ $\quad =$

12 $\frac{7}{10} + \frac{6}{10} + \frac{5}{10} =$ $\quad =$

13 $\frac{3}{5} + \frac{4}{5} + \frac{4}{5} =$ $\quad =$

14 $\frac{3}{10} + \frac{9}{10} + \frac{9}{10} =$ $\quad =$

15 $\frac{5}{8} + \frac{7}{8} + \frac{5}{8} =$ $\quad =$

SET 4 Extension

1 15.25 m = ☐ cm

2 How many months in 3.75 years?

3 What fraction of 2 m is 25 cm?

4 Add all even numbers between 31 and 37.

5 How many faces has a pentagonal pyramid?

6 Is 57 a prime number?

7

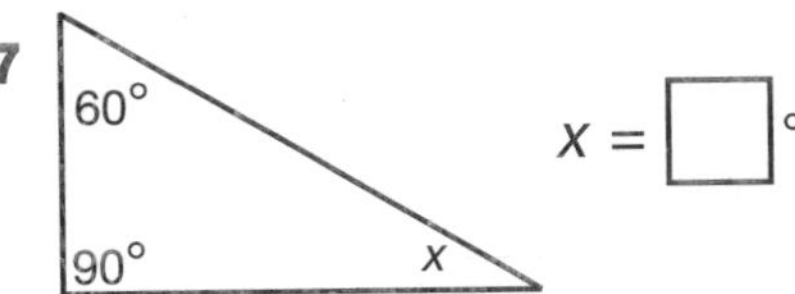

8 A hectare could be viewed as a square with 100 m sides. True or false?

9 If a 250 L tank is $\frac{7}{10}$ full, how much more could it hold?

10 Write 11:50 pm in 24-hour time.

11 Which is larger, 77.7 or $77\frac{7}{100}$?

12 Decrease 1 426 231 by 400 000.

13 7.1 L = ☐ mL

14 How many axes of symmetry has a kite?

15 What object does this net make?

16 How many 2 cm^3 cubes could fit into a box with length 8 cm, width 4 cm and height 10 cm?

Statistics and Probability Chance from zero to one

Rate the likelihood of spinning each colour on the spinner using the range 0–1.

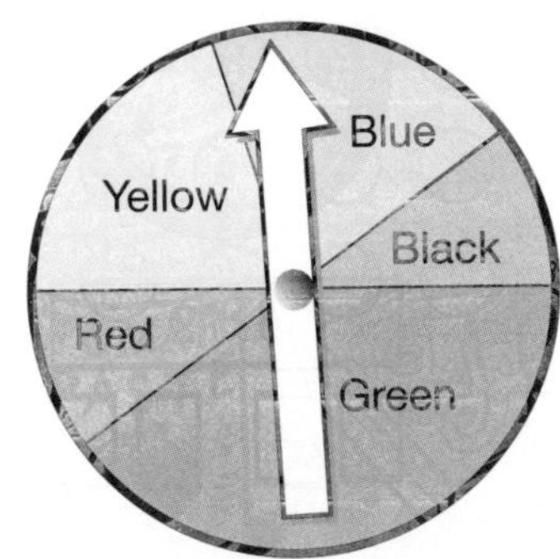

	Colour	Probability
1	Yellow	
2	Blue	
3	Black	
4	Green	
5	Red	

CERTAIN — 1
0.9
0.8
0.7
0.6
EVEN — 0.5
0.4
0.3
0.2
0.1
IMPOSSIBLE — 0

Number and Algebra

SET 1 Basic

1 6 × 8

2 7 × 9

3 60 – 15

4 16 ☐ 5 = 3 r 1

5 48 ÷ 8

6 22 ☐ 100 = 122

7 Is 17 a prime number?

8 Difference between 46 and 24

9 3 × 7 + 3 × 3

10 Factors of 18

11 \$37.45 = ☐ c

12 How many sides has a quadrilateral?

13 Minutes in $3\frac{1}{2}$ hours

14 7 m and 1 cm = ☐ cm

15 Value of 8 in 13 784

16 If the tickets cost \$28 each, how much will 3 tickets cost?

\$ ☐

SET 2 Subtracting decimals/money

1 Subtract 24 hundredths from 19.68.

2 17.68 – 9.95

3 25.08 – 13.17

4 Subtract 15 hundredths from 15.97.

5 Subtract 13 hundredths from 8.58.

6 \$276.48 – 64.99

7 \$595.83 – 49.09

8 23.424 – 5.398

9 44.698 – 15.907

10 Calculate each person's total wage after they received a bonus.

	A	B	C
Wage	\$235.15	\$587.20	\$678.90
Bonus	\$98.70	\$166.65	\$199.00
Total			

11 Calculate each person's take-home pay after tax was deducted.

	A	B	C
Wage	\$285.85	\$348.20	\$785.11
Tax	\$74.60	\$80.15	\$290.00
Take-home			

Measurement and Space Measuring angles

Measure the angles at the vertex of the intersecting lines.

1

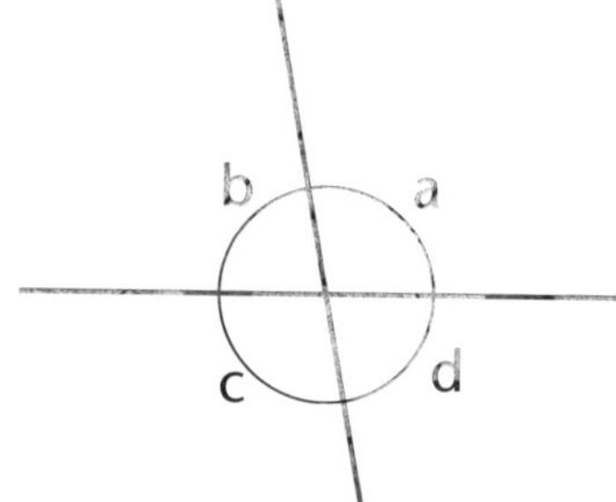

Angle	Degrees
a	°
b	°
c	°
d	°

2

Angle	Degrees
e	°
f	°
g	°
h	°

Number and Algebra

SET 3 Following rules

Complete the function of the machines.

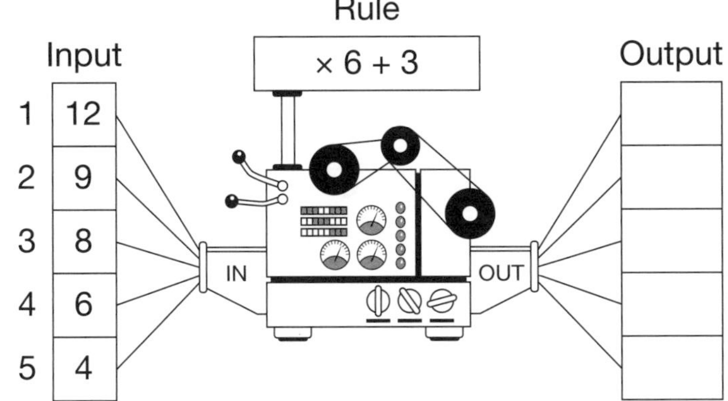

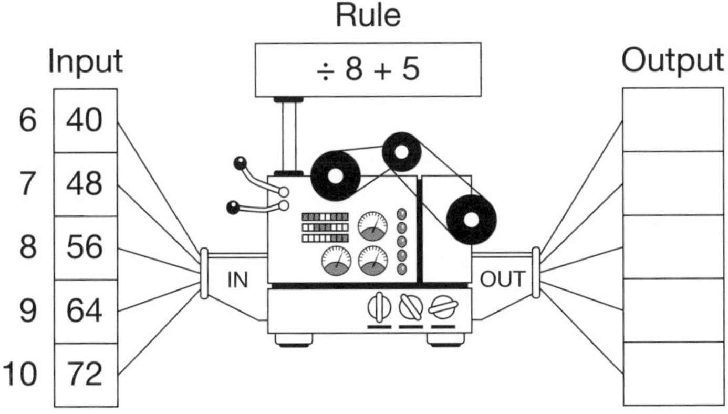

SET 4 Extension

1 $\frac{1}{4}$ of a day = ☐ minutes

2 0.25 of \$200

3 25% of \$36

4 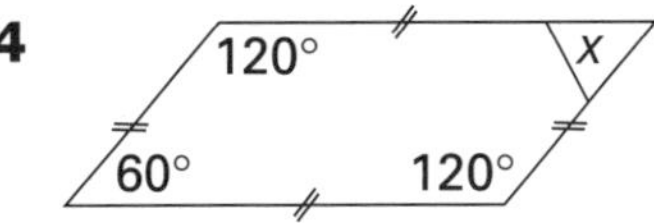

x = ☐ °

5 $7^2 - 4^2 + 37 - 16$

6 Value of 9 in 3 936 426

7 How many 750 mL buckets are needed to fill a 6 L container?

8 Gianni travelled 630 km in 7 hours. What was his average speed?

9 Days from 16 August to 17 September

10 Average of 4.5, 6.3, 7.2 and 6

11 Value of 6 in 6 342 937

12 Write 10:57 pm in 24-hour time.

13 How much is 8.75 kg at \$16 per kilogram?

14 How many axes of symmetry has a semi-circle?

15 Farel is 1.75 m tall. How tall is Joe if he is $\frac{4}{5}$ of Farel's height?

Measurement Volume and capacity

1 Complete the chart to show the relationship between volume and capacity.

Volume	100 cm^3		2500 cm^3	3000 cm^3	750 cm^2		2000 cm^3		1000 cm^3	2900 cm^3
Capacity	100 mL	500 mL				90 mL		2515 mL	1 L	

2 How much water will a 5 cm × 5 cm × 5 cm cube displace? ________

____

3 Jack placed a model into water and it displaced 1 litre of water. What was the volume of the model? ________________

____

UNIT 19

Number and Algebra

SET 1 Basic

1 7 × 8

2 95 – 36

3 Grams in 4 kg

4 33 + 22 + 11

5 63 ☐ 9 = 7

6 96 ÷ 8

7 120 ÷ 2

8 Divide 63 by 7.

9 $576.31 = ☐ c

10 $1.20 × 4

11 Value of 7 in 33 706

12 Are 18 and 42 multiples of 6?

13 Does a pentagon have 6 sides?

14 How many hours from 9 am to 2 pm?

15 How much change would I get from $2 if I bought nine 15c lollies?

$ 

SET 2 Dividing 5-digit numbers

1 $8\overline{)5208}$

2 $4\overline{)27\,636}$

3 $10\overline{)2734}$

4 $7\overline{)54\,439}$

5 $6\overline{)49\,087}$

6 $10\overline{)98\,700}$

7 $5\overline{)55\,271}$

8 $9\overline{)88\,888}$

9 Divide 6000 by 4.

10 Share $1800 among 6.

11 Pour 1305 mL evenly into 5 cups.

12 How many nines in 135?

13 1250 ÷ 10

14 ☐ ÷ 9 = 85

15 ☐ ÷ 10 = 91

16 412 ÷ ☐ = 103

17 5555 ÷ ☐ = 1111

Measurement and Space Perimeter

Use the grid paper to create 3 rectangles that have a perimeter of 24 cm.

Try to keep them apart from each other.

Number and Algebra

SET 3 Percentages

Calculate the saving on each item and the reduced price.

1

Runners	
Saving	
Price	

2

Boots	
Saving	
Price	

3

High heel sandals	
Saving	
Price	

4

Flip-flops	
Saving	
Price	

Calculate these percentages.

5 25% of 80 children

6 50% of $48

7 10% of 70 litres

8 30% of 100 trees

9 75% of 40 goals

10 20% of $80

Write a percentage for each fraction.

11 $\frac{1}{4}$ ☐

12 $\frac{4}{10}$ ☐

13 $\frac{1}{2}$ ☐

14 $\frac{1}{5}$ ☐

SET 4 Extension

1 2300 hrs = ☐ pm

2 What is the third angle of a triangle if the other angles are 48° and 29°?

3 How many hours in 3.75 days?

4 Which is greater, 25% of 1000 or 40% of 600?

5 What is the value of 9 in 28.194?

6 Round 158 960 to the nearest hundred.

7 A cook dropped $\frac{3}{8}$ of 4 dozen eggs. How many eggs were left?

8 Complete this sequence: 1, 4, 9, ☐, 25, ☐, 49

9 Estimate 298 × 21.

10 How much is 5.3 m at $7 per metre?

11 Reduce 258 008 by 10 000.

12 20% of $750

13 $\frac{24}{100} = \frac{\square}{25} =$

14 Which is larger, 510 mm or 0.5 m?

15 $\frac{1}{4} + \frac{1}{2} + \frac{7}{8} =$

16 Mr Flook had 400 sheep. If 20% of them died during a drought and 10% of the rest were sold, how many were left?

Measurement Kilograms and tonnes

How many of each set of masses are needed to make a tonne?

1

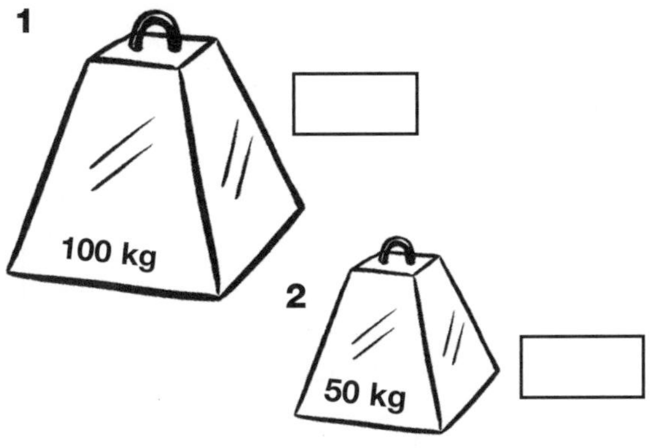

3

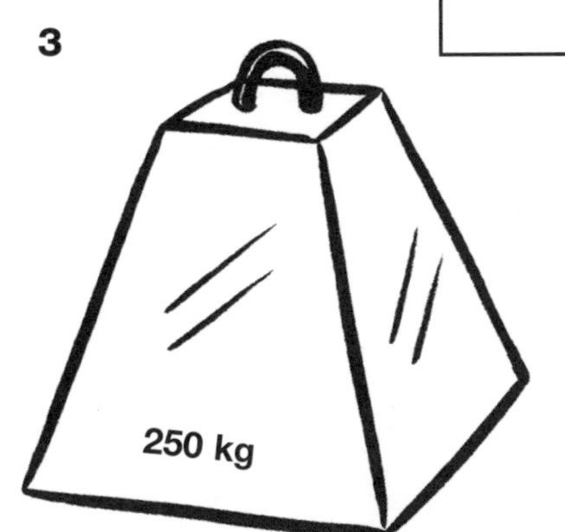

4

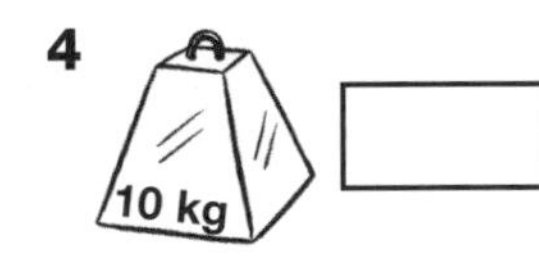

5

UNIT
20

Number and Algebra

SET 1 Basic

1 85 – 29

2 Degrees in half a circle

3 30 ÷ 6

4 72 ☐ 28 = 100

5 Subtract 38 from 49.

6 2 + 40 + 1000 + 600

7 How many 20c coins in $10?

8 Factors of 21

9 $10 – $3.75

10 Add 6 to the product of 3 and 2.

11 1997 + 35

12 Value of 6 in 9.06

13 121 ☐ 11 = 11

14 16 kg and 25 g = ☐ g

15

SET 2 Dividing by tens/averages

360 ÷ 40?
Think 360 ÷ 10 = 36
then divide that
answer by 4 to give 9.

1 $50\overline{)450}$

2 $80\overline{)560}$

3 $30\overline{)270}$

4 $40\overline{)320}$

5 $90\overline{)720}$

6 $60\overline{)3600}$

7 $70\overline{)630}$

8 $50\overline{)600}$

9 $80\overline{)3200}$

10 $40\overline{)2800}$

11 $70\overline{)4900}$

12 $60\overline{)5400}$

Find the average age of each group of children.

13

Average age ☐

14

Average age ☐

Space Reflections

Use the dot paper to reflect each shape.

Number and Algebra

SET 3 Square numbers

Join the dots to create sets of square numbers.

1 3^2

3 × 3 = ☐

2 4^2

4 × 4 = ☐

3 5^2

5 × 5 = ☐

SET 4 Extension

1 0.85 = ☐ %

2 $\frac{4}{10} + \frac{3}{10} + \frac{57}{100}$

3 How much are 24 stickers at 6 for 55c?

4 How much is 5.2 m of timber at $6 per metre?

5 How many minutes from 6:45 pm to 10:07 pm?

6 If a tap drips 2 L every 10 minutes, how many litres would it drip in a day?

7 Round 410 601 to the nearest thousand.

8 Add the prime numbers between 4 and 15.

9 How many litres of oil could I buy with $18 if oil costs $2.25 per litre?

10 20% of 90

11 If Mum bought 40 L of petrol and paid 85c per litre, how much did it cost her?

12 Suki's average batting score over 5 games was 16. Give a set of 5 scores that have a mean score of 16.

Working Mathematically

Statistics and Probability Sector graphs

Dawn spends one hour each night at swimming training.

True or false?

1 About 30 minutes is spent on freestyle. ☐

2 About 13 minutes is spent on backstroke. ☐

3 About 7 minutes is spent on butterfly. ☐

4 About 10 minutes is spent on breaststroke. ☐

5 About half her time is spent on backstroke, breaststroke and butterfly. ☐

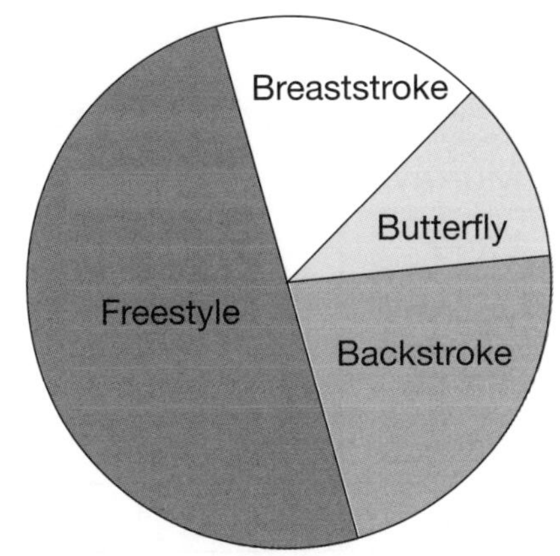

UNIT 21

Number and Algebra

SET 1 Basic

1 Half of 8^2
2 10:15 + 30 mins
3 420 – 170
4 How many quarters in $2\frac{1}{4}$?
5 (6 + 3) × 7
6 \$9.51 = ☐ c
7 10°C less than boiling point
8 2.1 × 7
9 2.4 + 3.5
10 94 + 72 + 6
11 Which is larger, 5^2 or 30?
12 Product of 9 and 9
13 29 km + 38 km
14 Difference between 3 and 40
15

SET 2 Extended multiplication

1 468 × 35

2 170 × 42

3 521 × 62

4 471 × 38

5 342 × 83

6 612 × 65

7 365 × 54

8 197 × 39

9 604 × 71

Working Mathematically

Circle the mistakes in these algorithms.

10
```
      4 2 1
  ×     3 6
  ---------
    2 4 2 6
  1 2 6 3 0
  1 5 0 5 6
```

11
```
      6 4 9
  ×     2 8
  ---------
    5 1 9 7
  1 2 8 8 0
  1 8 0 7 7
```

Statistics and Probability Two-way tables

Complete the totals for each column on the cricket scoresheet to show how many overs the bowlers have bowled, how many wickets they have taken and how many runs the opposition batters have hit off their bowling.

Bowler	Overs	Wickets	Runs
Murray	8	2	16
Lane	7	0	34
Vanda	9	4	28
Simms	6	1	22
Wallace	5	3	20
Total			

1 Which bowler conceded the most runs?
2 Who was the most successful wicket taker?
3 Did the bowler who bowled the most overs take the most wickets?
4 Did the bowler who bowled the least overs concede the least number of runs?
5 Do you agree that on average the team takes one wicket every 12 runs?

Number and Algebra

SET 3 Add and subtract fractions

Record these answers as improper fractions and as mixed numerals.

1 $\frac{8}{10} + \frac{9}{10} = \quad =$

2 $\frac{2}{4} + \frac{3}{4} = \quad =$

3 $\frac{3}{4} + \frac{3}{4} + \frac{1}{4} = \quad =$

4 $\frac{7}{10} + \frac{6}{10} + \frac{5}{10} = \quad =$

5 $\frac{3}{5} + \frac{4}{5} + \frac{4}{5} = \quad =$

Subtract the fraction from the whole numbers

6 $1 - \frac{2}{3} =$

7 $3 - \frac{1}{2} =$

8 $4 - \frac{3}{4} =$

9 $5 - \frac{1}{8} =$

10 $4 - \frac{1}{3} =$

Add or subtract these mixed numerals

11 $4\frac{1}{3} + 2\frac{1}{3} =$

12 $5\frac{7}{8} - 2\frac{3}{8} =$

13 $4\frac{4}{5} - 2\frac{1}{5} =$

SET 4 Extension

1 Round and estimate 18.9×4.3.

2 $8.42 × 10

3 Average of 270, 360 and 540

4 15% of $90

5 Estimate an answer to 39×306.

6 72.4 + 49.27

7 $\frac{1}{4}$ kg at $5.20 a kg

8 Order 137%, 1.4, $1\frac{3}{10}$ and 1.45.

9 $(4.5 + 4.5) \times 6$

10 Round 826 900 to the nearest 1000.

11 Write 127% as a decimal.

12 How many seconds in 3 hours?

13 Sum of 3^2, 4^2 and 5^2

14 Add the prime numbers between 40 and 50.

Working Mathematically

15 How many 4 cm^3 cubes would fit inside this prism? ________

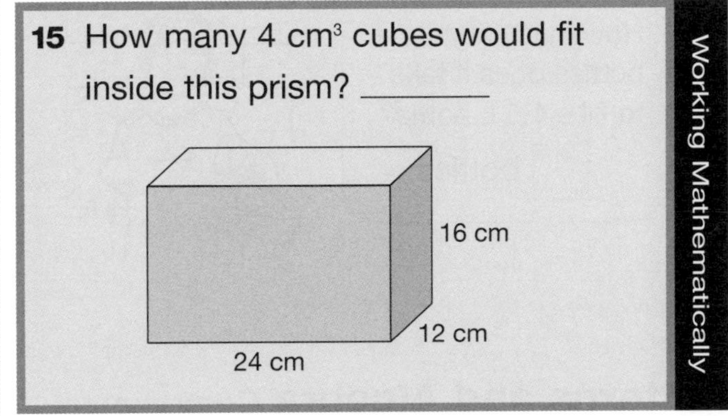

Measurement Elapsed time

Draw the time each person finished work on the clock faces.

Ben started work at 9:00 am and finished $9\frac{1}{2}$ hours later.	Jessica started work at 8:15 am and finished $10\frac{1}{2}$ hours later.	Keera started work at 7:45 am and finished 9 hours and 8 minutes later.	Denzel the night worker started at 9:30 pm and worked for 11 hours and 22 minutes.
1	2	3	4

UNIT
22

Number and Algebra

SET 1 Basic

1 34 ÷ 6 = 5 r ☐
2 450 – 160
3 1 + 10 + 500 + 6000
4 (5 + 4) × 3
5 2.1 + 4.8
6 190 – 55
7 $\frac{1}{4}$ kg at $4.20 per kg
8 9 × 7
9 3.2 × 3
10 Quotient of 63 and 9
11 2.5 – 1.6
12 46 ÷ 5 = 9 r ☐
13 25 m + 38 m
14 Product of 8 and 12
15

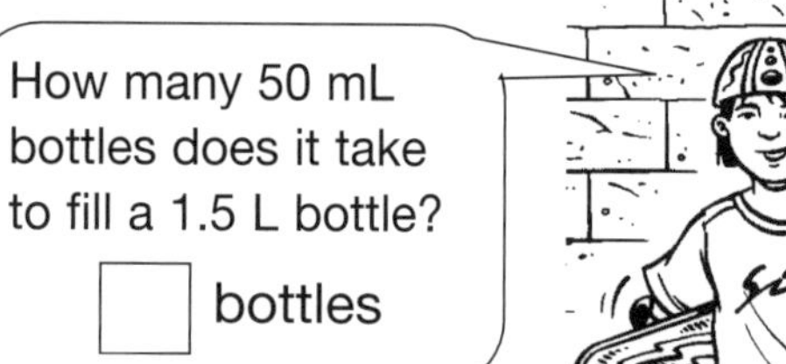

SET 2 Geometric patterns

Complete each table to show the number of sides needed to make each set of 2D shapes.

1

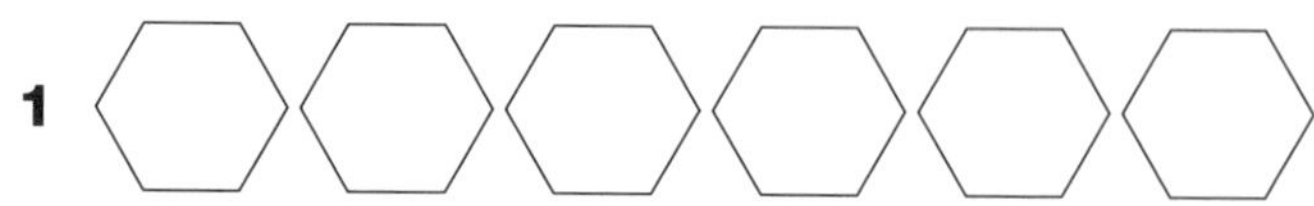

Hexagons	1	2	3	4	5	6
Sides						

2

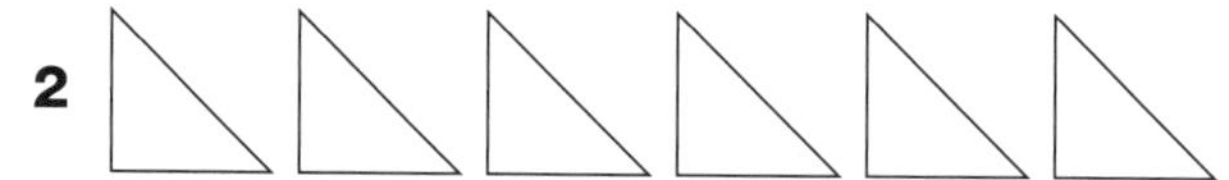

Triangles	1	2	3	4	5	6
Sides						

3

Octagons	1	2	3	4	5	6
Sides						

4

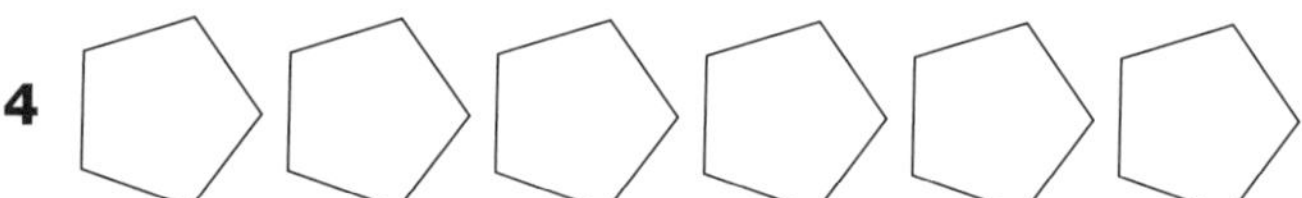

Pentagons	1	2	3	4	5	6
Sides						

Patterns and Algebra Coordinates

Plot the coordinates and then connect them to create shapes.

1

(0,0)
(2,2)
(2,5)
(3,6)
(4,5)
(4,2)
(6,0)

2

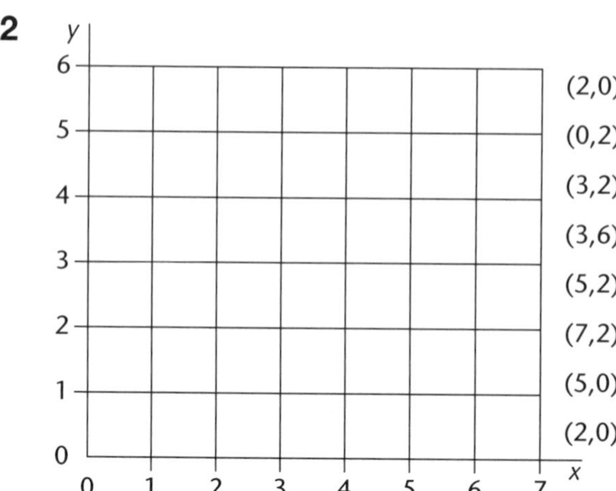

(2,0)
(0,2)
(3,2)
(3,6)
(5,2)
(7,2)
(5,0)
(2,0)

Number and Algebra

SET 3 Making equivalent fractions

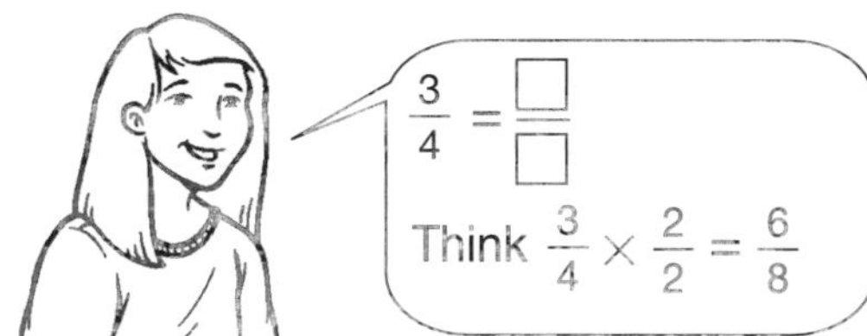

Make an equivalent fraction by multiplying the numerator and denominator by the same number.

1 $\frac{1}{5} \times \frac{2}{2} =$

2 $\frac{1}{10} \times \frac{3}{3} =$

3 $\frac{1}{6} \times \frac{4}{4} =$

4 $\frac{2}{5} \times \frac{3}{3} =$

5 $\frac{3}{4} \times \frac{5}{5} =$

6 $\frac{2}{3} \times \frac{4}{4} =$

Divide the numerator and denominator by the same number to make equivalent fractions.

7 $\frac{4}{10} \div \frac{2}{2} =$

8 $\frac{9}{15} \div \frac{3}{3} =$

9 $\frac{8}{12} \div \frac{4}{4} =$

10 $\frac{10}{15} \div \frac{5}{5} =$

11 $\frac{6}{10} \div \frac{2}{2} =$

SET 4 Extension

1 Write 9:27 am in 24-hour time.

2 3.75 km = $\square$ m

3 Write the prime numbers between 21 and 33.

4 Average of \$1.25, \$3.75 and \$1.45

5 $\frac{\square}{4} = \frac{30}{40}$

6 How many degrees in 3 triangles?

7 Write fifty-five thousand, two hundred and seventeen in figures.

8 How many 1500 mL buckets are needed to fill a 24 L can?

9 What are the sides of a square if the area is 25 cm²?

10 How many 175 mL bottles can be filled from 1.575 L?

11 Area of a rectangle with sides of 24 cm and 20 cm

12 A \$280.00 item at 20% off

13 What object does this net make?

14 How much is 4.7 kg at \$2 per kg?

15 | 90° | x |
|---|---|
| 90° | 90° |

$x = \square$°

Statistics and Probability Interpreting data/line graphs

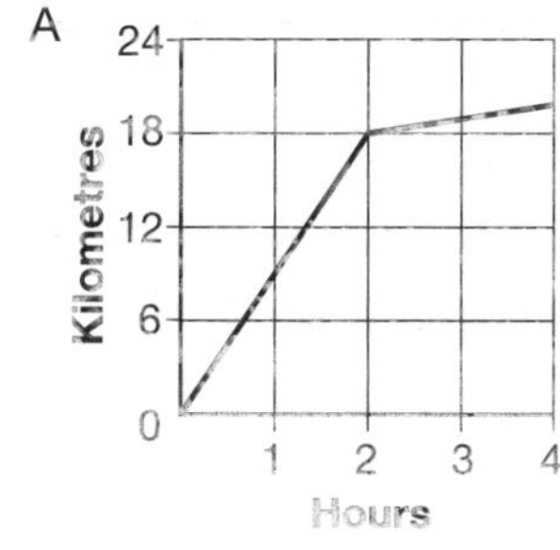

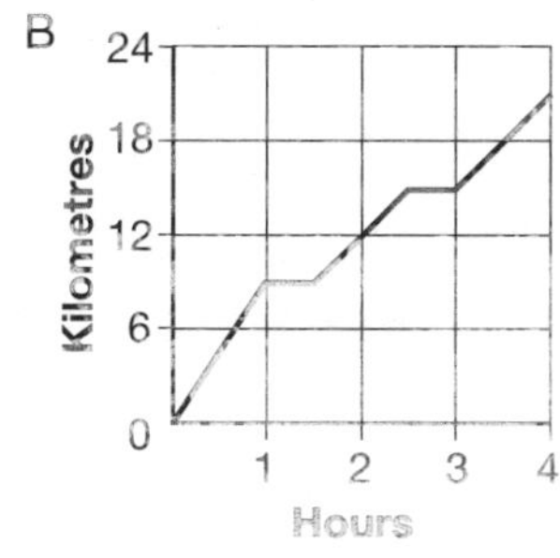

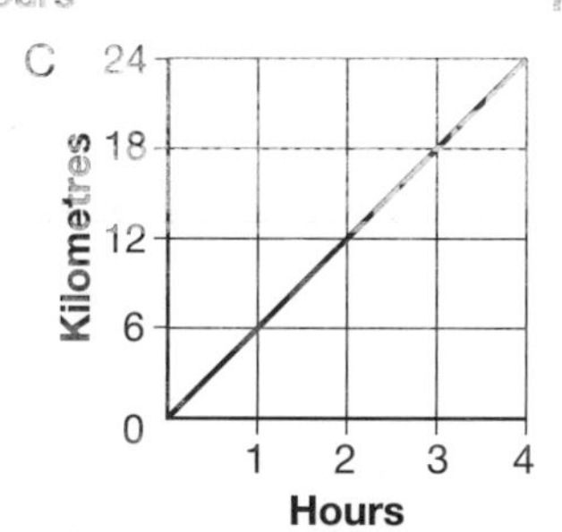

Study the graphs that show how three people performed in a 24 km fun run. Answer true or false.

1 After 1 hour A was at the 9 km mark. $\square$

2 After 1 hour C had travelled 6 km. $\square$

3 B had a 30-minute rest after 1 hour. $\square$

4 A averaged 6 km/h for the first 2 hours. $\square$

5 B averaged less than 6 km/h for the whole race. $\square$

6 Slow and steady wins the race. $\square$

UNIT 23

Number and Algebra

SET 1 Basic

1 8 + 9 + 13

2 1000 ÷ 2

3 1000 – 600

4 1000 – 50

5 42 × 10

6 60 ☐ 2 = 30

7 8 ☐ 10 = 80

8 $3 \times 8 + 3^2$

9 Write the factors for 30.

10 Is 49 a prime number?

11 23 215, 23 230, 23 245, ☐

12 2 m and 97 cm = ☐ cm

13 5 hundreds + 2420

14 How many hours from 11 am to 3 pm?

15 How many minutes in $\frac{1}{3}$ of an hour? ☐ minutes

SET 2 Multiplying decimals

1 13.2 × 4

2 22.4 × 5

3 20.7 × 6

4 14.32 × 5

5 10.08 × 7

6 25.86 × 8

7 213.7 × 3

8 224.07 × 6

9 243.82 × 5

10 Five girls need 1.75 m of material each to make a costume for play night. How much material will they need altogether?

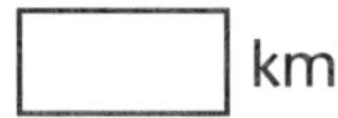

☐ m

11 A farmer has a large property that measures 7 km by 8.08 km. What is its area?

☐ km^2

Space Coordinates

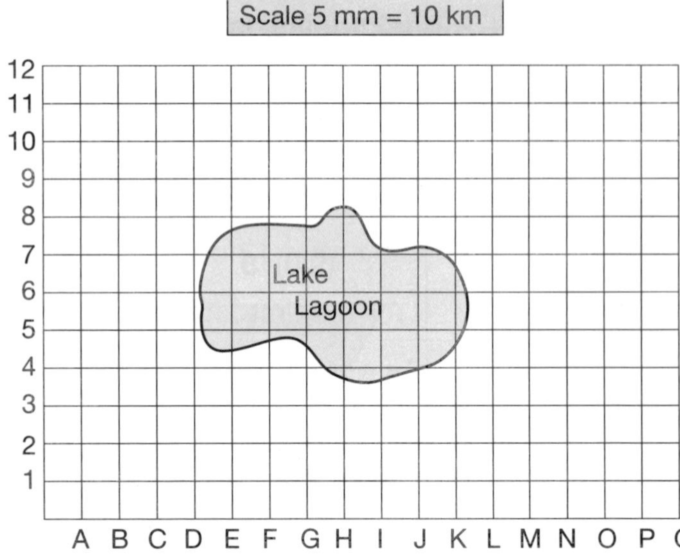

Put a dot on the map for each person's house.

1 Kelly (A,1)

2 Ben (A,10)

3 Taryn (N,10)

4 Jim (N,3)

5 Fred (F,3)

6 Lauren (P,6)

Use the scale to calculate the distance between:

7 Kelly's house and Ben's house.

8 Ben's house and Taryn's house.

9 Taryn's house and Jim's house.

Number and Algebra

SET 3 Equivalent fractions

1 whole									
$\frac{1}{2}$	$\frac{1}{2}$								
$\frac{1}{4}$	$\frac{1}{4}$	$\frac{1}{4}$	$\frac{1}{4}$						
$\frac{1}{8}$	$\frac{1}{8}$	$\frac{1}{8}$	$\frac{1}{8}$	$\frac{1}{8}$	$\frac{1}{8}$	$\frac{1}{8}$	$\frac{1}{8}$		
$\frac{1}{5}$	$\frac{1}{5}$	$\frac{1}{5}$	$\frac{1}{5}$	$\frac{1}{5}$					
$\frac{1}{10}$	$\frac{1}{10}$	$\frac{1}{10}$	$\frac{1}{10}$	$\frac{1}{10}$	$\frac{1}{10}$	$\frac{1}{10}$	$\frac{1}{10}$	$\frac{1}{10}$	$\frac{1}{10}$
$\frac{1}{3}$	$\frac{2}{3}$	$\frac{3}{3}$							
$\frac{1}{6}$	$\frac{2}{6}$	$\frac{3}{6}$	$\frac{4}{6}$	$\frac{5}{6}$	$\frac{6}{6}$				

Write an equivalent fraction for each.

1 $\frac{2}{5} = \frac{\square}{10}$

2 $\frac{3}{4} = \frac{\square}{8}$

3 $\frac{1}{3} = \frac{\square}{6}$

4 $\frac{1}{2} = \frac{\square}{8}$

5 $\frac{2}{3} = \frac{\square}{6}$

6 $\frac{8}{10} = \frac{\square}{5}$

7 $\frac{2}{8} = \frac{\square}{4}$

8 $\frac{4}{6} = \frac{\square}{3}$

9 $\frac{6}{8} = \frac{\square}{4}$

10 $\frac{3}{6} = \frac{\square}{2}$

True or false?

11 $\frac{1}{5} > \frac{3}{10}$

12 $\frac{5}{8} < \frac{3}{4}$

13 $\frac{5}{10} = \frac{3}{6}$

14 $\frac{2}{3} > \frac{5}{6}$

SET 4 Extension

1 Write 10:15 pm in 24-hour time.

2 Order 3.3, 3.19, $3\frac{1}{5}$ and $3\frac{1}{3}$.

3 20% of $1500

4 What fraction of 3 kg is 600 g?

5 $10^3 + 10^2 + 10 + 1$

6 How much is 6.7 kg at $5 per kg?

7 Reduce 180 by 40%.

8 Subtract the product of 9 and 4 from 9^2.

9 I was paid $15 per hour and I worked from 8:45 am till 12:45 pm. How much did I earn?

10 48 minutes before 21:15

11 Difference between 27.6°C and 34.1°C

12 I went to bed at 10:30 pm and slept for $8\frac{1}{4}$ hours. When did I get up?

13 There are 12 rows of 108 seats. How many seats are there altogether?

14 Average of 8.2, 9.4, 3.5 and 2.9

15 What is the perimeter of this trapezium?

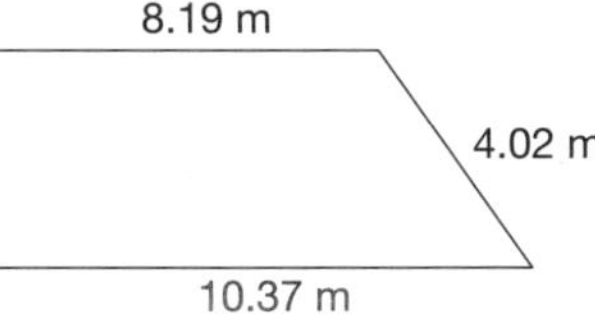

Measurement and Space Quadrilaterals

1 Choose the correct name from the word bank to name each shape.

square	rectangle	rhombus	trapezium

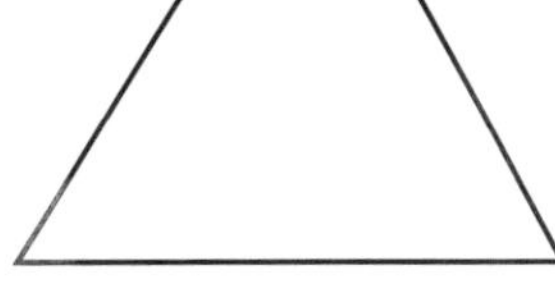

__________ __________ __________ __________

2 Which of the above shapes are also parallelograms?

Number and Algebra

SET 1 Basic

1 8 + 14 + 6
2 Is 17 a prime number?
3 18 ☐ 2 = 36
4 19 × 10
5 1900 − ☐ = 1000
6 64 ÷ 8
7 Tens in 5460
8 Factors of 27
9 $8.15 × 100
10 85 × 100
11 Average of 6, 9 and 12
12 72 ÷ 9
13 Hundreds in 5289
14 Value of 8 in 85 962
15 150 cm × 3
16

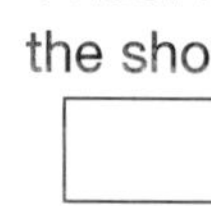

The show starts at 10:15 and finishes at 11:30. How long is the show?

☐ minutes

SET 2 Dividing large numbers

Use your calculator to crack the code.

452	258.9	86.5	1458	15.8	12.5	14.5
N	E	M	I	L	U	W

1 $9\overline{)4068}$

2 $10\overline{)2589}$

3 $23\overline{)333.5}$

4 $28\overline{)2422}$

5 $8\overline{)11\,664}$

6 $33\overline{)521.4}$

7 $55\overline{)869}$

8 $3\overline{)776.7}$

9 $22\overline{)9944}$

10 $20\overline{)9040}$

11 $17\overline{)24\,786}$

12 $114\overline{)1425}$

13 $52\overline{)4498}$

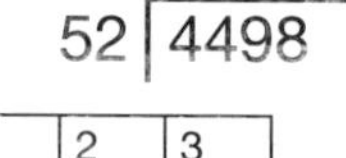

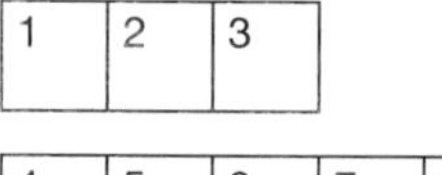

Space Adjacent angles

Calculate the size of each angle by subtracting the given angle from 180°.

1

2
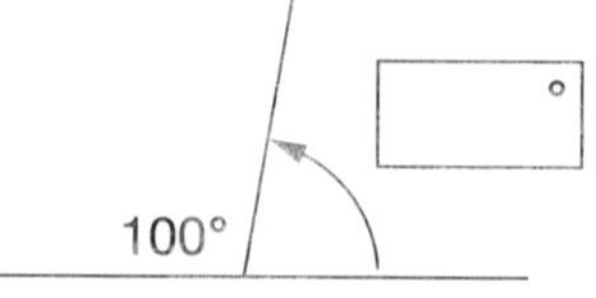

3
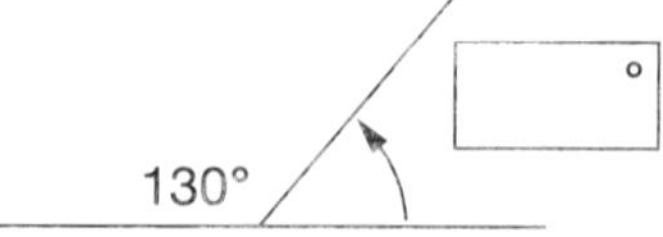

Calculate the size of each reflex angle by subtracting the given angle from 360°.

4
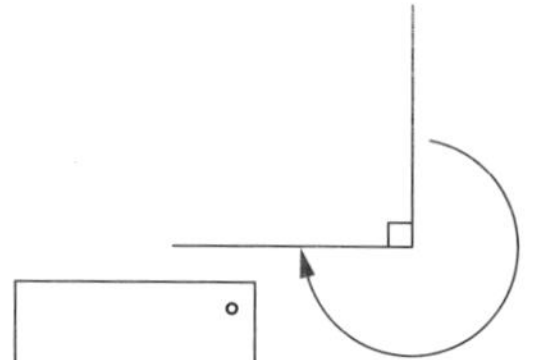

5
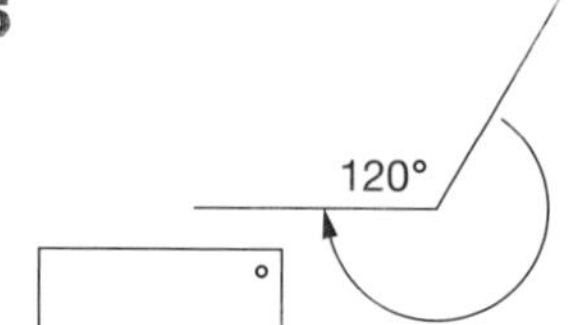

6
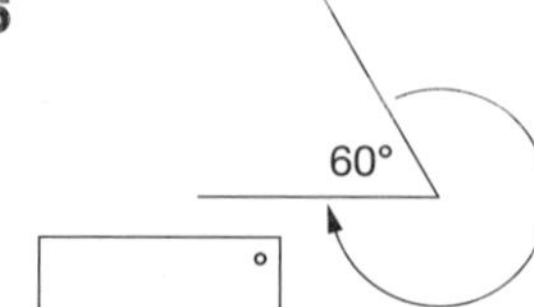

Number and Algebra

SET 3 Add and subtract fractions with related denominators

1 whole

$\frac{1}{2}$ | $\frac{1}{2}$

$\frac{1}{4}$ | $\frac{1}{4}$ | $\frac{1}{4}$ | $\frac{1}{4}$

$\frac{1}{8}$ | $\frac{1}{8}$ | $\frac{1}{8}$ | $\frac{1}{8}$ | $\frac{1}{8}$ | $\frac{1}{8}$ | $\frac{1}{8}$ | $\frac{1}{8}$

$\frac{1}{5}$ | $\frac{1}{5}$ | $\frac{1}{5}$ | $\frac{1}{5}$ | $\frac{1}{5}$

$\frac{1}{10}$ | $\frac{1}{10}$ | $\frac{1}{10}$ | $\frac{1}{10}$ | $\frac{1}{10}$ | $\frac{1}{10}$ | $\frac{1}{10}$ | $\frac{1}{10}$ | $\frac{1}{10}$ | $\frac{1}{10}$

1 $\frac{1}{2} + \frac{1}{4} =$

2 $\frac{1}{4} + \frac{1}{8} =$

3 $\frac{1}{5} + \frac{1}{10} =$

4 $\frac{1}{2} + \frac{1}{10} =$

5 $\frac{1}{4} + \frac{2}{8} =$

6 $3\frac{1}{5} + 1\frac{2}{10} =$

7 $4\frac{5}{10} + 2\frac{1}{5} =$

8 $6\frac{3}{8} + 1\frac{1}{4} =$

9 $\frac{1}{2} + \frac{2}{10} =$

10 $\frac{4}{5} - \frac{2}{10} =$

11 $\frac{1}{2} - \frac{1}{10} =$

12 $\frac{5}{8} - \frac{1}{4} =$

13 $\frac{9}{10} - \frac{1}{2} =$

14 $8\frac{3}{4} - 1\frac{1}{2} =$

15 $9\frac{7}{8} - 2\frac{1}{2} =$

16 $7\frac{4}{5} - 3\frac{4}{10} =$

SET 4 Extension

1 7.45 – 3.23

2 Which is not equivalent, $\frac{2}{5}$, 0.35, 0.4 or 40%?

3 Average of 3.7, 4.5, 4.3 and 3.5

4 $7 \times 20 + 1$

5 $\frac{8}{10} - \frac{71}{100}$

6 $1 - \frac{2}{100}$

7 $6^2 + 3^2$

8 $\frac{35}{5} = \frac{\square}{10}$

9 How many 375 mL soft drink cans are needed to fill a 3 L bucket?

10 How many thousands in 236 107?

11

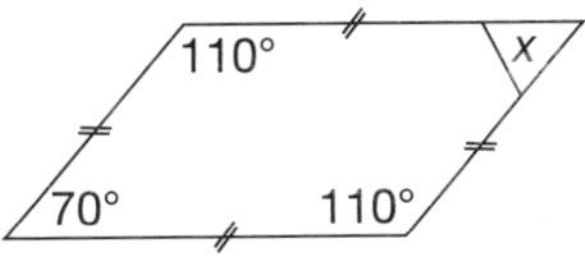

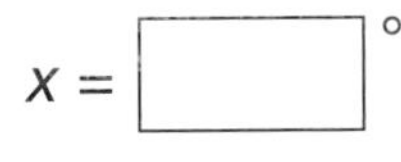

Working Mathematically

Calculate the difference in hours and minutes between these times.

12 1550 and 5:30 pm ____________

13 1415 and 4:50 pm ____________

14 2110 and 11:45 pm ____________

15 1750 and 7:05 pm ____________

Measurement The cubic centimetre

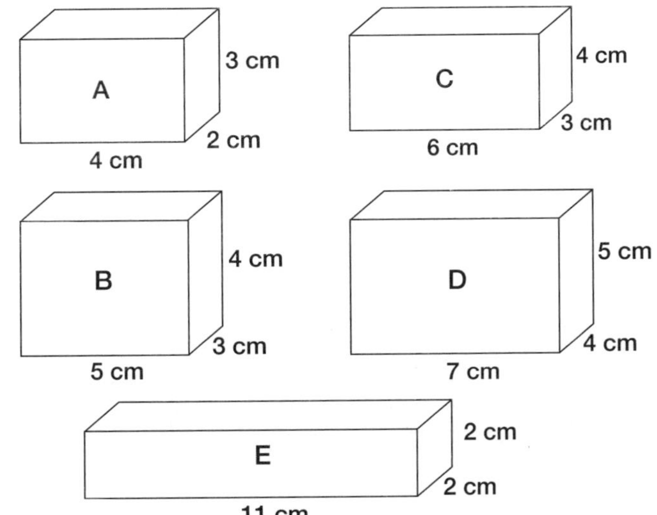

Calculate the volume of these prisms.

	Shape	Length	Width	Height	Volume
1	A				
2	B				
3	C				
4	D				
5	E				

UNIT 25

Number and Algebra

SET 1 Basic

1 7 ☐ 23 = 30

2 7×9

3 13c × 100 = $ ☐

4 $\frac{2}{10} + \frac{5}{10}$

5 5 + 25

6 80 ☐ 50 = 30

7 63 ☐ 9 = 7

8 Halves in $1\frac{1}{2}$

9 0.3×2

10 $3 \times 7 + 3^2$

11 Is 36 a multiple of 5?

12 Is 35 a prime number?

13 235 306, 235 311, 235 316, ☐

14 How many minutes in $\frac{1}{2}$ hour?

15 How much change would I get from $50 if I bought 8 chocolates at $3 each?
$ ☐

SET 2 Addition of 5- and 6-digit numbers

THE COLONIAL
$120 795

VISTA
$354 999

SEAHAVEN
$215 605

Calculate these additions.

1 The Coopers are building the Colonial on a block of land costing $172 500. The total cost is $ ☐

2 The Rose family built the Vista on a block of land costing $545 800. They spent $15 620 on landscaping. The total cost is $ ☐

3 The Evans are building the Seahaven and a pool worth $16 200 on land worth $77 550. The total cost is $ ☐

Space Reflect, translate, rotate

1 Rotate 90° clockwise

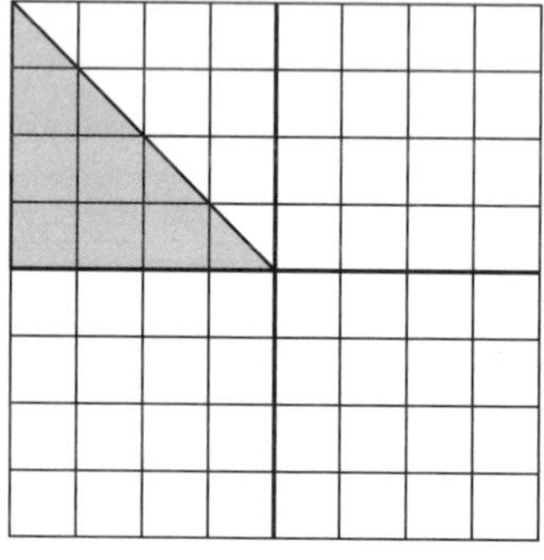

2 Translate directly below

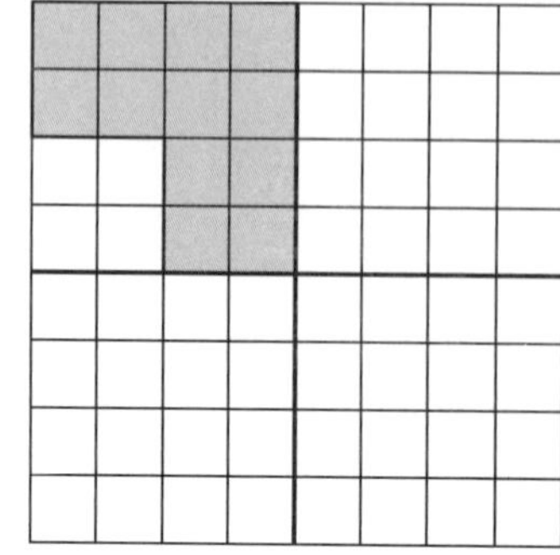

3 Rotate 180° clockwise

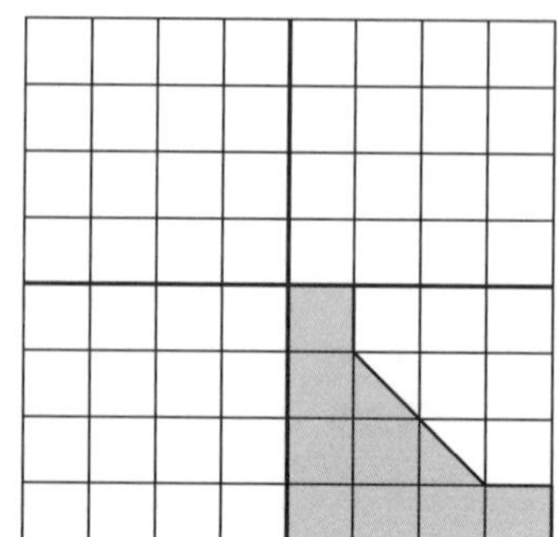

Number and Algebra

SET 3 Multiplication by tens

1

	×10
7	
11	
26	
35	
20	

2

	×10
80	
90	
100	
127	
316	

3

	×100
8	
16	
23	
100	
118	

4

	×1000
6	
15	
50	
100	
225	

Working Mathematically

Calculate the price of 100 items if you know how much 10 cost.

	Item	Cost of 10	Cost of 100
5	Football	$175	
6	Racquet	$238	
7	Bat	$334	

SET 4 Extension

1 $6^2 + 4^2$

2 100 + 5 × 4 + 79

3 Add the difference between 180 and 130 to 70.

4 30 + 20 × 4 + 5

5 60 – 80 ÷ 5 + 6

6 Which is the smallest: $\frac{7}{10}$, 75% or 0.71?

7 2.6 × 10 + 4

8 42.7 ÷ 7

9 25% of $280

10 How much is 4500 g at $8.80 a kg?

11 1 m + 25 cm + 4 mm = ☐ mm

12 Round off then estimate 18.9 × 5.2.

13 How much interest is earned in one year if $800 is invested at 10% p.a.?

14 Complete this sequence: 144, 121, ☐, ☐, ☐, 29

15 Factors of 42

16 A batsman hit 12 fours, 2 sixes and 47 singles. How many runs did he score?

17 Luis is $\frac{4}{5}$ of Finn's height. If Finn is 1.8 m, how much shorter is Luis?

Statistics and Probability Sample data

30 children selected at random were used as a sample group to identify the most popular flavours of ice creams.

Use the data to predict how many of each flavour should be ordered to cater for 150 children at an end of year picnic.

Flavours	Vanilla	Chocolate	Strawberry	Mango	Rainbow
Sample group (30 children)	6	10	9	2	3
Whole school (150 children)					

UNIT 26

Number and Algebra

SET 1 Basic

1 58c × 10

2 Factors of 45

3 0.4 × 6

4 $\frac{4}{10} + \frac{5}{10}$

5 Product of 9 and 12

6 How much is 250 g at $4.80 a kg?

7 3.4 × 4

8 Is 37 a multiple of 7?

9 Quotient of 81 and 9

10 4.2 × 10 + 18

11 $6^2 + 6$

12 59 ÷ 6

13 10:25 – 19 mins

14 Difference between 10^2 and 9^2

15

SET 2 Multiplication (4 digits × 2 digits)

1 2363×38

2 3145×86

3 4185×27

4 2503×79

5 4327×51

6 3385×18

Working Mathematically

7 The hall held 27 rows of 38 seats and all the seats had been sold for the concert. A ticket cost $23. What was the total amount collected?

8 There are 48 bags each containing 26 lollipops. If each lollipop weighs 20 g, what is the total weight of the 48 bags?

Number and Algebra Rounding numbers

Round each number to the nearest 10, 100 and 1000.

	Number	Nearest 10	Nearest 100	Nearest 1000
1	3592			
2	1055			
3	3529			
4	873			
5	9318			
6	6666			

Round each number to the nearest 1000 to estimate an answer, then calculate the exact answer.

	Question	Estimate	Exact
7	1891 + 4163	2000 + 4000 = 6000	6054
8	8761 – 2845		
9	6117 + 3922		
10	3618 + 2386		
11	7762 + 1224		
12	5681 – 3749		

Number and Algebra

SET 3 Negative numbers

The vertical number line at the side may help you complete this.

1	$3 + 5 =$	6	$4 - 8 + 2 =$
2	$3 - 5 =$	7	$2 - 9 + 3 =$
3	$-3 + 5 =$	8	$1 - 5 + 7 =$
4	$-4 + 7 =$	9	$-8 + 1 - 4 =$
5	$2 - 6 =$	10	$-7 + 2 + 3 =$

8 7 6 5 4 3 2 1 0 –1 –2 –3 –4 –5 –6 –7 –8

Create number sentences to solve the problems.

11 What was Willow's final score in the indoor cricket game if her scores were 9, –18 and 7?

12 What is Noah's final bank balance for this month if he deposited and withdrew these amounts +$54, –$20 and –$45?

13 What was Harper's final score on the television game show if her scores were 25, –10, –10 and 15?

SET 4 Extension

1 What is the difference in height between Kimberley (131 cm) and Jacqueline (191 cm)?

2 Perimeter of a rectangular field 65 m wide and 110 m long

3 Divide 240 m into 4 equal lots.

4 6 L of paint at $19.99 a litre

5 How many minutes in 12.7 hours?

6 Difference between 1 hr 14 min and 95 min

7 $\frac{3}{5} + \frac{2}{5} + \frac{7}{10} + \frac{1}{10}$

8 0.75 L × 100

9 60% of $150

10 $9.90 less 10%

11 How many whole litres of petrol could I buy with $28 if it cost 87c per litre?

12 Stuart has read $\frac{4}{5}$ of his book. What percentage has he still to read?

13 How many 275 mL jars can be filled from 2.2 L?

14 $\frac{8}{10}$ of 3 L

15 An intravenous drip releases 3 mL of fluid every 20 seconds. How much will it release in $1\frac{3}{4}$ hours?

Statistics and Probability Data surveys

Write 3 survey questions you could use to collect data to find out how students want to improve their school playground, e.g. Should we have more grass areas?

1 ____________________

2 ____________________

3 ____________________

Number and Algebra

SET 1 Basic

1 49 + 17
2 $6^2 + 18$
3 (19 – 11) × 7
4 25 – 6 + 1
5 Add 15 minutes to 10:55 am.
6 Subtract 7^2 from 100.
7 72c × 100
8 Factors of 21
9 Find the sum of odd numbers less than 10.
10 Find the sum of even numbers less than 10.
11 105 × 10
12 1000 ÷ 10
13 37° – 9°
14 \$5.00 – \$3.78
15

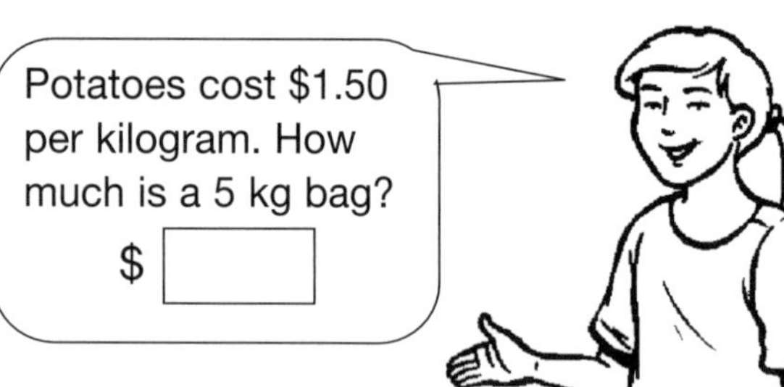

SET 2 Fraction and decimal remainders

Solve these divisions, writing your remainder as a fraction.

1 $3\overline{)16}$ = $5\frac{1}{3}$
2 $4\overline{)25}$
3 $5\overline{)256}$
4 $4\overline{)343}$
5 $3\overline{)742}$
6 $4\overline{)345}$
7 $5\overline{)297}$
8 $6\overline{)787}$
9 $7\overline{)2256}$
10 $8\overline{)3468}$
11 $9\overline{)6598}$
12 $10\overline{)3593}$

Solve these divisions, writing your remainder as a decimal. You may need a calculator.

13 $4\overline{)333}$ = 83.25
14 $2\overline{)397}$
15 $4\overline{)505}$
16 $4\overline{)747}$
17 $10\overline{)3897}$
18 $10\overline{)2679}$
19 $5\overline{)366}$
20 $5\overline{)5767}$
21 $5\overline{)5378}$

Space The Cartesian plane

Plot these ordered pairs on the part of the Cartesian plane supplied.

The first one is done for you.

1 (5,7)
2 (4,6)
3 (3,5)
4 (2,4)
5 (1,3)
6 (0,2)
7 (–1,1)
8 (–2,0)
9 Draw a line to join the dots.

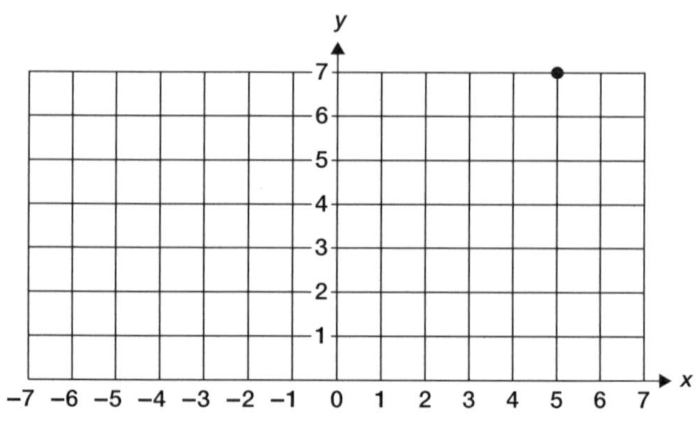

Number and Algebra

SET 3 Using fractions to record division

1 Share 5 between 4.

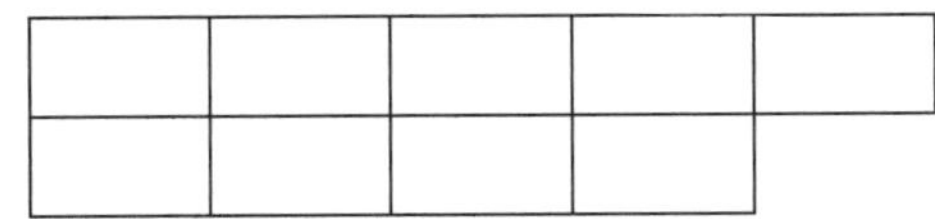

$5 \div 4$ becomes $\frac{5}{4}$ = _______

2 Share 6 between 5.

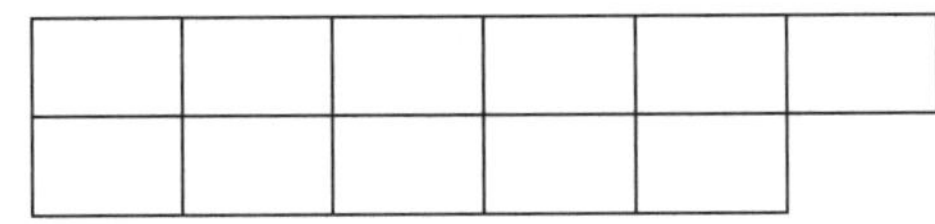

$6 \div 5$ becomes $\frac{6}{5}$ = _______

3 Share 3 between 2.

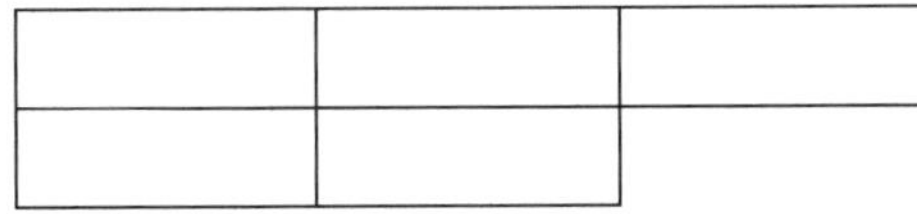

$3 \div 2$ becomes $\frac{3}{2}$ = _______

4 Share 4 between 3.

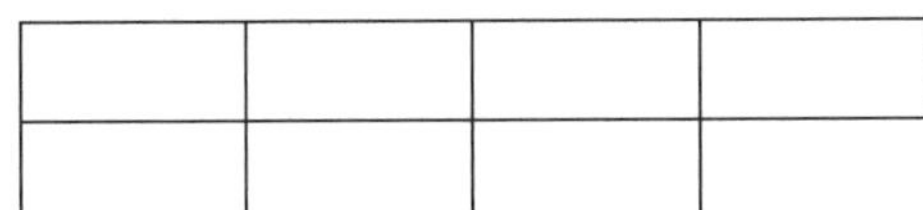

$4 \div 3$ becomes $\frac{4}{3}$ = _______

SET 4 Extension

1 What is the perimeter of a quadrilateral with sides of 7.6 m, 3.4 m, 4.1 m and 5.2 m?

2 Which does not belong: 90%, $\frac{9}{10}$, $\frac{91}{100}$ or 0.9?

3 How many combinations of clothing can be made from 3 shirts and 3 pairs of pants?

4

110° x 90°   60°

x = ☐°

5 Change $\frac{43}{5}$ to a mixed number.

6 Round 49 899 to the nearest 1000.

7 Dad's car covers 12 km per litre. If petrol costs $0.85 per litre, how much will a trip of 408 km cost?

8 Create a question that matches this working out.

$\overset{168}{7 \times 24} \times 60 =$ _______

____

____

____

Working Mathematically

Measurement Hectares

A hectare can be looked upon as a square block of land with 100 m sides. How many blocks of land of each shape would fit into a hectare?

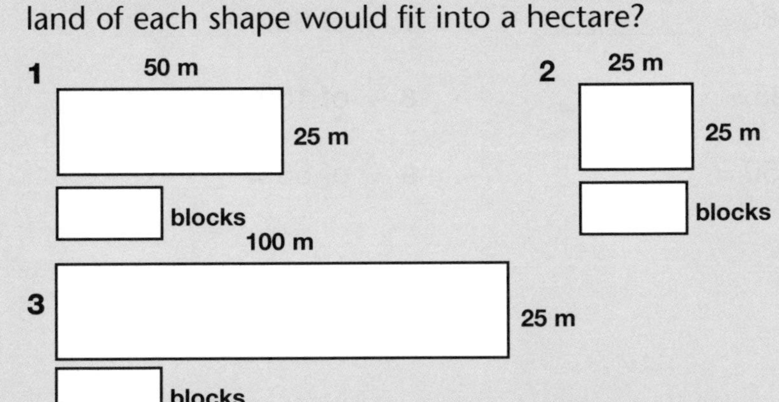

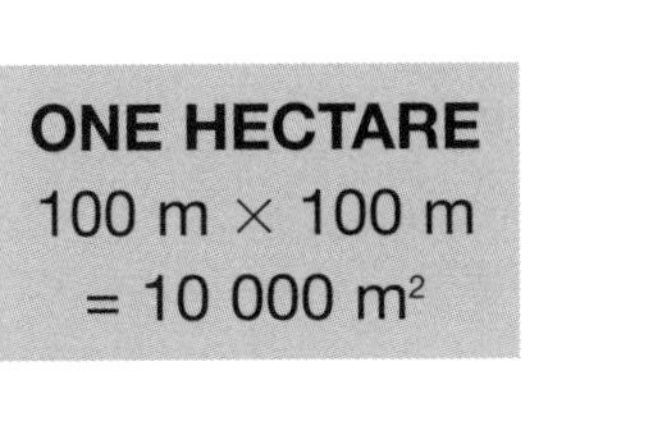

Working Mathematically

Number and Algebra

SET 1 Basic

1 25 ☐ 75 = 100

2 9 × 9

3 400 ☐ 250 = 150

4 18 + 24

5 80 ☐ 10 = 8

6 36 – 23

7 13 ☐ 3 = 39

8 $17.45 = ☐ c

9 4526 + 3000

10 How much change from $15 would I receive if I spent $7.50?

11 How much are 5 books at $1.40 each?

12 How many $\frac{1}{4}$s in 2?

13 How many tens in 150?

14 44 + ☐ = 5 × 10

15

SET 2 Calculators

FRESH FRUIT SHOP	
Bananas $3.50 per kg	Grapes $4.20 per kg
Pears $2.50 per kg	Tomatoes $2.70 per kg

Use your calculator to work out the total of these bills.

Alison's bill

1	4 kg of tomatoes	
2	2 kg of pears	
3	2 kg of bananas	
4	1 kg of grapes	
5	Total	

Brendan's bill

6	$2\frac{1}{2}$ kg of grapes	
7	3 kg of bananas	
8	$1\frac{1}{2}$ kg of pears	
9	1 kg of tomatoes	
10	Total	

11 Alison's change from $50 ☐

12 Brendan's change from $50 ☐

13 Total mass of Alison's groceries ☐

14 Total mass of Brendan's groceries ☐

Number and Algebra Fractional quantities

1 $\frac{3}{5}$ of 40 = ________

2 $\frac{1}{6}$ of 36 = ________

3 $\frac{3}{8}$ of 24 = ________

4 $\frac{3}{10}$ of 60 = ________

5 $\frac{4}{5}$ of 35 = ________

6 $\frac{6}{10}$ of 50 = ________

7 $\frac{3}{4}$ of 60 = ________

8 $\frac{1}{5}$ of 100 = ________

9 $\frac{5}{8}$ of 56 = ________

Number and Algebra

SET 3 Positive and negative numbers

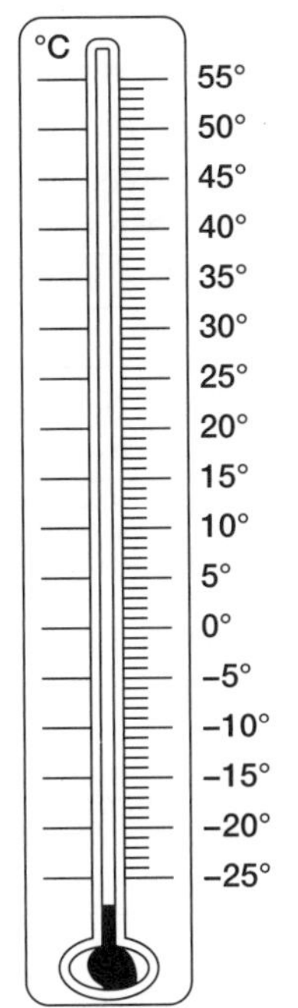

Use the thermometer to solve the problem.

1. 10°C + 25°C
2. 10°C – 15°C
3. –10°C + 15°C
4. –10°C + 25°C
5. –25°C + 30°C
6. –15°C + 25°C
7. 10°C – 25°C
8. 5°C – 20°C
9. 15°C – 25°C
10. –5°C + 25°C

11. At midnight the temperature was –5°C. By 6 am it had risen by 5°C. At noon it had risen by another 10°C. By 6 pm the temperature had dropped 5°C. What was the temperature at 6 pm?
12. If the temperature was –15°C at 4 am and rose by an average of 5°C every 4 hours, what would the temperature be at 4 pm?

SET 4 Extension

1. Round 85 961 to the nearest hundred.
2. Value of 9 in 4.8907
3. 65% of 100 + 10% of 200
4. Twenty-eight minutes later than 9:56
5. 85% = [0.   ]
6. $17 + \frac{1}{2} \times 96 - 33$
7. 45% of $900
8. 64.48 ÷ 8
9. Quotient of 69 and 9
10. What is the area of a square with a perimeter of 20 cm?
11. 85 minutes after 2345
12. 900 m + 20 m + 3 km = [   ] km
13. How many degrees in 2 circles?
14. 90% of $250
15. Order 375%, 3.7, $3\frac{8}{10}$ and 3.79.
16. If 7 books cost $63.49, how much for 9?

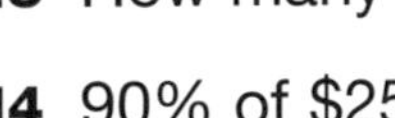

Measurement Timelines

Match the events in Peta's life to a place on the timeline.

Peta's timeline

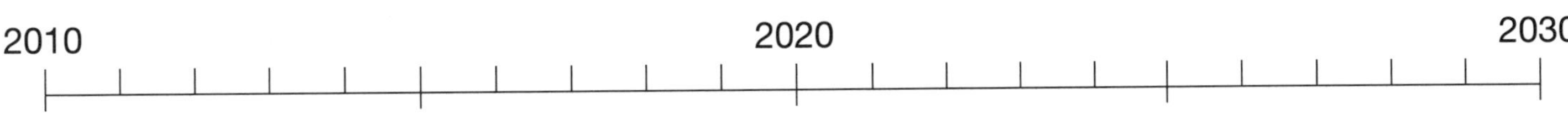

Started walking January 2012	**Started pre-school** June 2014	**Started school** February 2016	**Got the mumps** January 2023	**Made the softball team** February 2021	**Baby brother born** June 2022

UNIT 29

Number and Algebra

SET 1 Basic

1 Difference between 1000 and 20

2 Product of 8 and 20

3 $1.65 × 10

4 3.4 + 6.2

5 8 – 1.5

6 0.25 kg = ☐ g

7 How many hundreds in 5105?

8 How much is 2 L at $1.30 per litre?

9 Triple 25.

10 Divide 71 by 8.

11 $\frac{90}{100} - \frac{4}{10}$

12 98 + 150 + 2

13 10c × 1000

14 4 ×  = 1000

15

There were 200 questions and my mark was 60%. How many questions did I get right?

☐ questions

SET 2 Multiplication by 2 digits

Surf Supplies Incorporated 1998

Surfboard $399 | Wetsuit $250 | Fins $180

Rash shirt $79 | Zinc $2.50 | Wax $3.80

Calculate the income from the business last month if these sales were made.

	Item	Quantity	Cost per unit	Income
1	Board	38		
2	Wetsuit	29		
3	Fins	40		
4	Shirt	63		
5	Zinc	99		
6	Wax	95		

Calculate the total cost of these items.

7 2 boards and a shirt ______________

8 Fins, wax and zinc ______________

9 Wetsuit, fins and board ______________

10 5 shirts and 3 zincs ______________

Measurement and Space Volume

Record the length, width and height of each model in the grid. Use the formula ***Volume = length × width × height*** to calculate the volume of each model.

Model	Length cm	Width cm	Height cm	Volume cm^3
A				
B				
C				

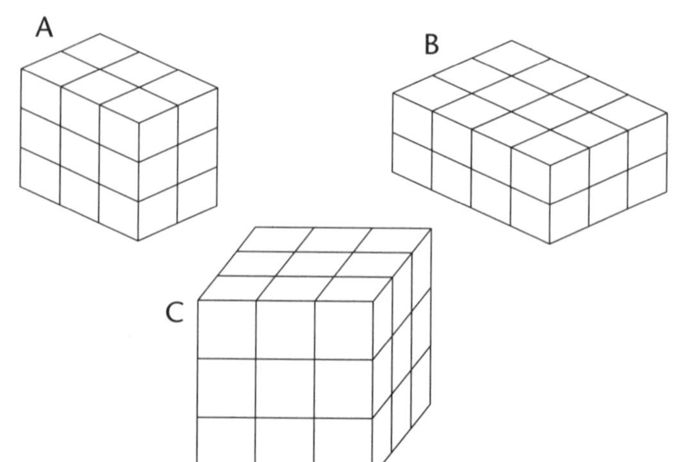

Number and Algebra

SET 3 Finding percentages

1 10% of $30

2 10% of 40 matches

3 20% of 30 pens

4 20% of 20 dogs

5 25% of 40 fish

6 25% of $60

7 50% of 24 sheep

8 20% of 60 pencils

200 spectators watched a cricket match. Calculate the following numbers of people.

9 75% wore hats

10 90% wore sunglasses

11 60% had soft drinks

12 50% bought pies

13 30% cheered for the winners

14 75% drove to the match

SET 4 Extension

1 11.11, 11.31, 11.51, $\square$

2 How many 1.35 m lengths can be cut from 6.75 m?

3 Write 9% as a decimal.

4 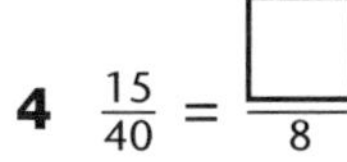 $\frac{15}{40} = \frac{\square}{8}$

5 Name this shape: 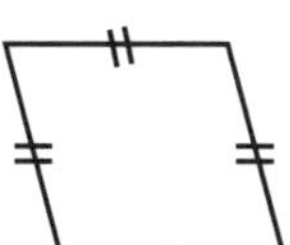

6 3.21 m = $\square$ cm

7 How many edges has a cube?

8 Perimeter of a square with an area of 64 cm^2

9 x / 60° $x = \square°$

Working Mathematically

Australia's population is 27 000 000.

Use a calculator to work out how many people belong to each age group.

	Age group	Number
10	21% are under 15	
11	67% are aged 15–64	
12	12% are 65 and over	

Number and Algebra Multiplication problems

1	4 litres of paint were used to paint a 24 m^2 wall. How many litres would be needed to cover a 96 m^2 surface?	24 m^2 / 4 L | | |
2	The 36 kg fruit box contained 4 watermelons. What would be the mass of 32 similar watermelons?	36 kg / 4 | | | | | | | |
3	Keirra received $15 for working 3 hours. How much would she receive if she worked 21 hours?	$15 / 3 hrs | | | | | |

UNIT 30

Number and Algebra

SET 1 Basic

1 How many mm in 1 m?

2 91 × 10

3 $4^2 + 5^2$

4 36 = $\square^2$

5 700 – 350

6 12 ☐ 5 = 60

7 36 ☐ 14 = 50

8 21 ÷ 5

9 Divide 60 by 5.

10 6 × 8 + 12

11 Difference between 270 and 50

12 Add 37 and 137.

13 69 + 30 + 1

14 $\frac{1}{3}$ of 15

15

Melissa poured 2 L and 350 mL into a jug. How many mL were there altogether?

☐ mL

SET 2 Decimal number patterns

Complete the function machines.

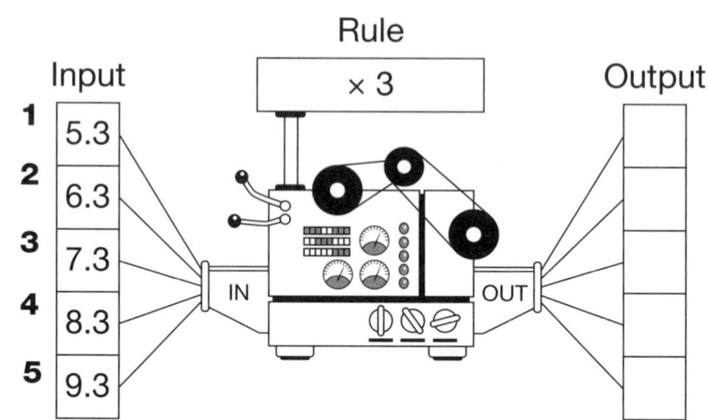

	Input	Output
1	5.3	
2	6.3	
3	7.3	
4	8.3	
5	9.3	

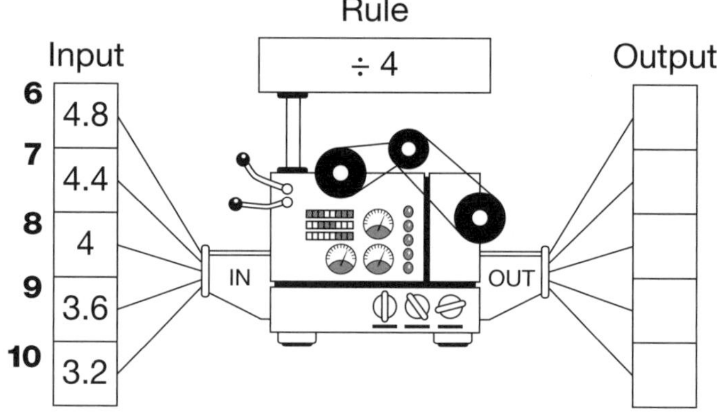

	Input	Output
6	4.8	
7	4.4	
8	4	
9	3.6	
10	3.2	

Measurement Choosing measuring devices

List 3 things that could be measured by each device.

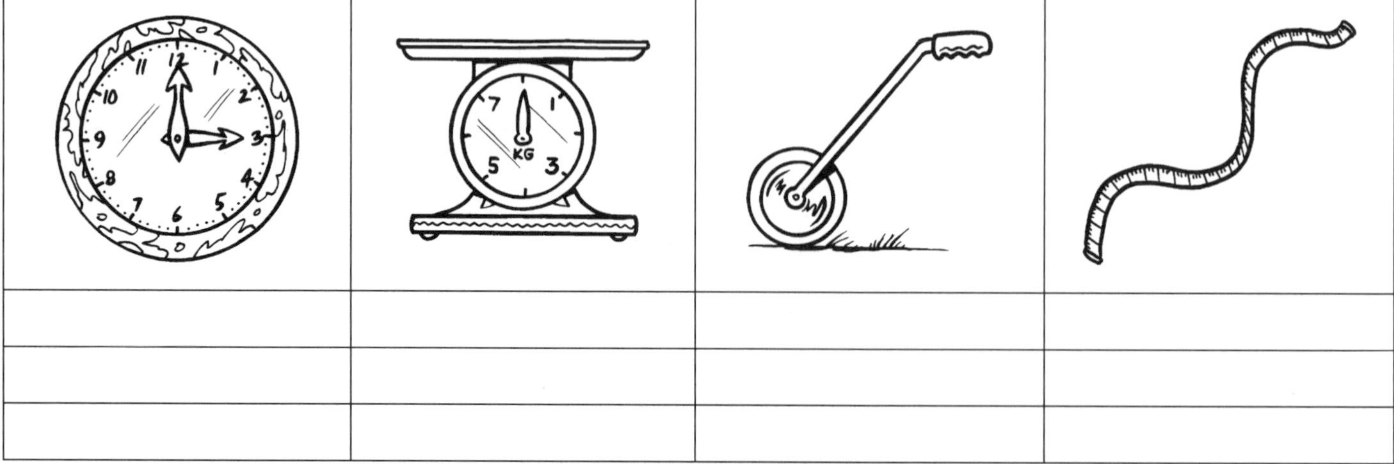

Number and Algebra

SET 3 Prime factors

Complete the factor trees.

1

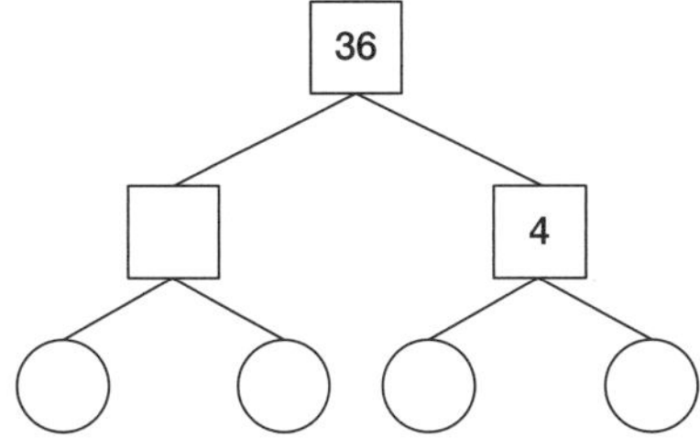

2

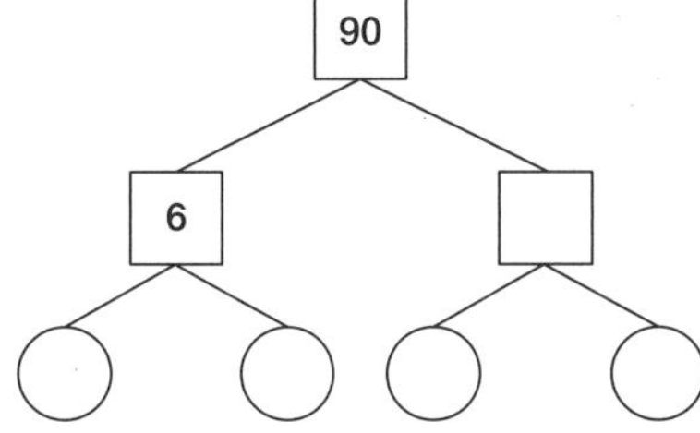

3

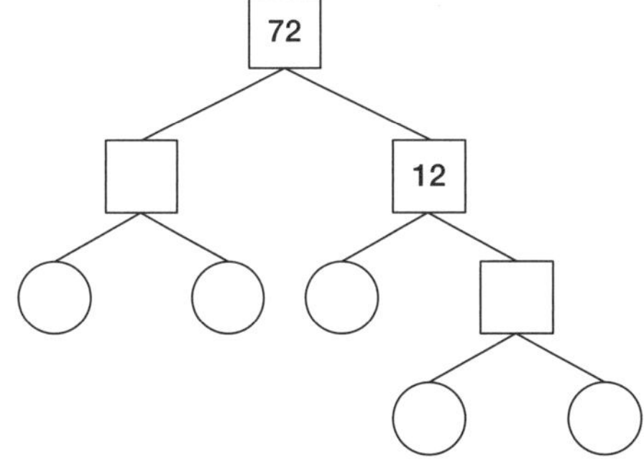

SET 4 Extension

1 5 m + 85 cm = ☐ cm

2 How much is 3 kg at $48 for 6 kg?

3 74% = ☐ hundredths

4 25% of $2600

5 Round and estimate 69.8 × 8.9.

6 3 × 60 ÷ 9 + 7

7 What fraction of 250 is 75?

8 56.16 ÷ 8

9 2 m + 32 cm + 8 mm = ☐ mm

10 Quotient of 774 and 8

11 How many minutes between 9:55 am and 2:24 pm?

12 How much is 750 g at $3.60 a kg?

13 What is the volume of a swimming pool with dimensions 9 m, 5 m and 2 m?

14 Round 1 987 621 to the nearest million.

15 800 m + 100 m + 2 km = ☐ km

16 Mia won $350 in Lotto. If she kept $\frac{3}{5}$ of it and gave the rest away, how much did she keep? $ ☐

Statistics and Probability The mean

Find the mean of these cricket scores.

	Player	Scores	Mean
1	Peter	10, 7, 17, 10, 5, 11	
2	Jessica	17, 17, 17, 21, 8	
3	Eli	30, 32, 26, 22, 30	
4	Selene	29, 35, 28, 35, 13	
5	Nico	3, 7, 6, 13, 6, 17, 4, 8	
6	Bruce	11, 15, 14, 21, 14, 25, 12, 16	

UNIT 31

Number and Algebra

SET 1 Basic

1 350 + 60

2 800 ☐ 2 = 400

3 Product of 6 and 10

4 Sum of 30 and 170

5 3 × 55c

6 Divide 43 by 6.

7 3 thousands + 17 231

8 Value of 9 in 923 231

9 Are 42 and 37 factors of 9?

10 List primes between 20 and 30.

11 Write the factors of 21.

12 $\frac{23}{100}$ = 0.☐

13 $1 - \frac{3}{4}$

14 $37.45 ☐ c

15

SET 2 Multiply and divide decimals

You may need to use your calculator for some of these questions.

	×	10	100	1000
1	0.231			
2	4.38			
3	0.643			
4	1.87			
5	15.6			

	÷	10	100	1000
6	3568			
7	4295			
8	235.2			
9	68.9			
10	19.5			

Write true or false.

11 3.68 × 10 = 36.8 ________

12 0.183 × 100 = 18.3 ________

13 23.6 ÷ 100 = 0.236 ________

14 18.5 ÷ 1000 = 0.0185 ________

15 2.81 ÷ 10 = 28.1 ________

16 0.392 × 1000 = 392 ________

Measurement and Space Transforming shapes

Reflect these 2D shapes along the dotted line to make new shapes.

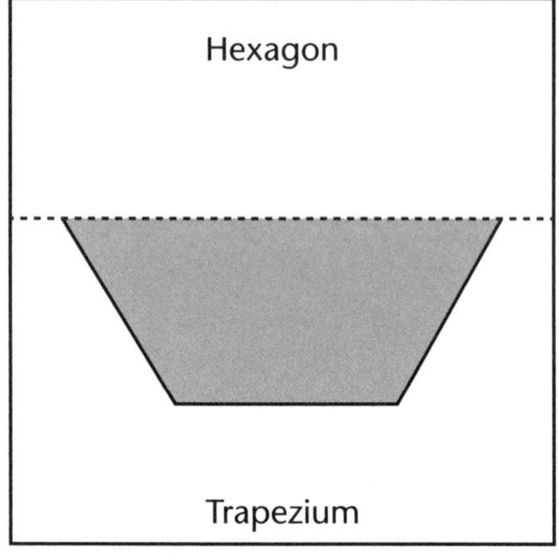

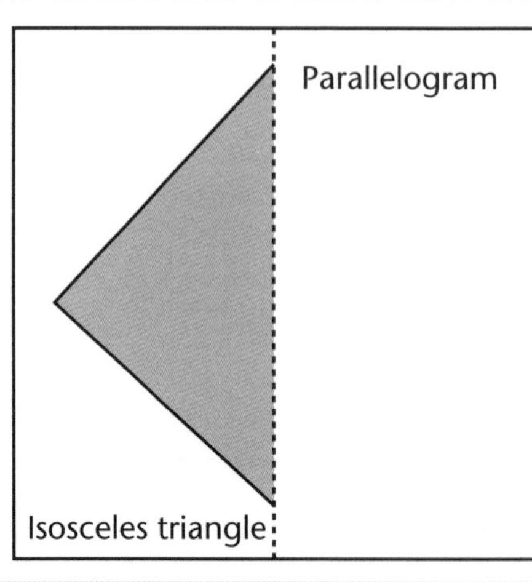

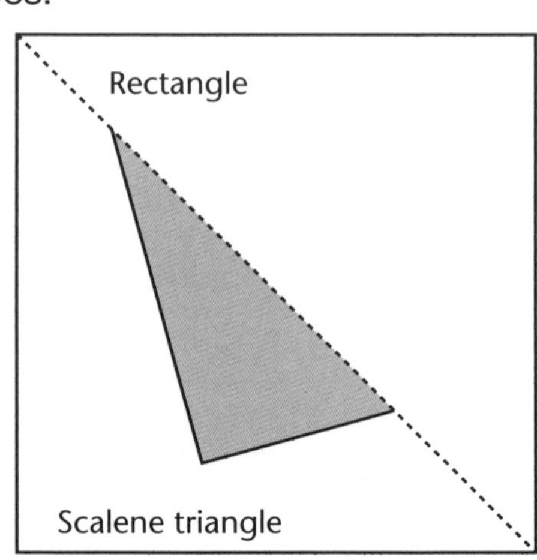

Patterns and Algebra

SET 3 Number patterns

Follow the rules to complete the grids.

1 ■ = ▲ × 8

▲	7	6	5	4	3	2	1	0
■								

2 ★ = ◆ × 6

◆	3	6	9	12	15	18	21
★							

3 ▲ + 99 = ■

▲	11	12	13	14	15	16	17
■							

4 ▲ ÷ 2 = ■

▲	200	180	160	140	120	100	80
■							

5 Create a pattern and write a rule to explain it.

▲	1	2	3	4	5	6	7
■							

Rule: ______________________

Working Mathematically

SET 4 Extension

1 5×10^2 + 18 tens

2 1, 2, 3, 5, 8, 13, ☐

3 Average of 11.4, 12.3, 9.3 and 3.4

4 500 × 341 × 2

5 3.5 ha ÷ 7

6 9.5 ÷ 5 × ☐ = 11.4

7 Area of a square that has a 28 cm perimeter

8 $\frac{76}{100} - \frac{5}{10}$

9 Centimetres in 1 km

10 How many minutes are there from 9:48 pm to 1:05 am?

11 $800 less 40%

12 2 m = 1500 mm + ☐ cm

13 What time is it $\frac{3}{4}$ of an hour before ten to five?

14 What is the radius of a circle that is 186 mm wide?

15 4.5 km minus $\frac{8}{10}$ of a kilometre

16 How many cubes measuring 3 cm by 3 cm by 3 cm can fit inside a box measuring 9 cm by 6 cm by 12 cm?

Measurement Kilometres

Calculate the kilometres travelled in these 4WD tours.

1 Start at **A** and drive to **D**, passing through **B** and **C**.

2 Start at **A** and drive to **D**, passing through **G** and **E**.

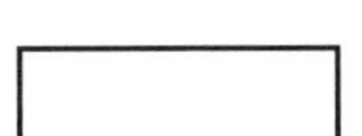

3 Start at **A** and drive to **D**, passing through **B**, **F** and **E**.

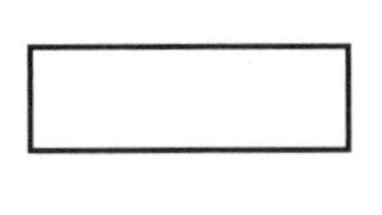

4 Start at **E** and drive to **D**, passing through **F**, **B** and **C**.

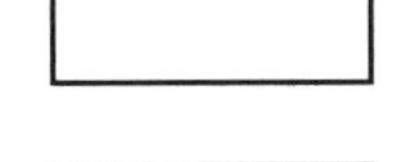

5 Start and end at **A**, passing through **B**, **C**, **D**, **E** and **G**.

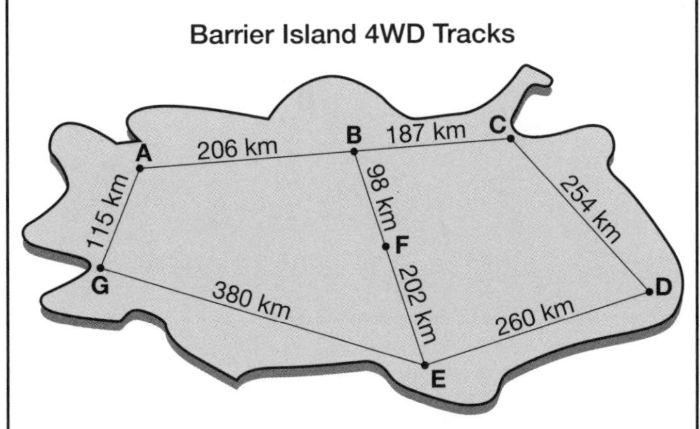

UNIT 32

Number and Algebra

SET 1 Basic

1 9^2

2 76c + 94c

3 80 × 5

4 Product of 9 and 3

5 40 ÷ 6

6 \$28 × 100

7 \$39 ÷ 3

8 0.65 = ☐ %

9 4 × ☐ = 448

10 7.7 + 2.13

11 0.8 + 0.8 + 0.8 + 0.8

12 Difference between 90 and 16

13 1 hectare = ☐ m^2

14 Value of 6 in 27.615

15 

How many minutes are there from 9:15 am to 11:07 am?

☐ minutes

SET 2 Decimal place value

1	×10	×100
0.7		
0.6		
0.3		
3.5		
4.8		

2	×10	×100
2.6		
5.8		
3.12		
4.08		
9.17		

3	×10	×100	×1000
28.52			
15.37			
18.24			
15.163			
36.202			
72.374			
51.727			

Space Cartesian plane

Write the ordered pairs for each letter.

1 A = (,)

2 B = (,)

3 C = (,)

4 D = (,)

5 E = (,)

6 F = (,)

7 G = (,)

8 H = (,)

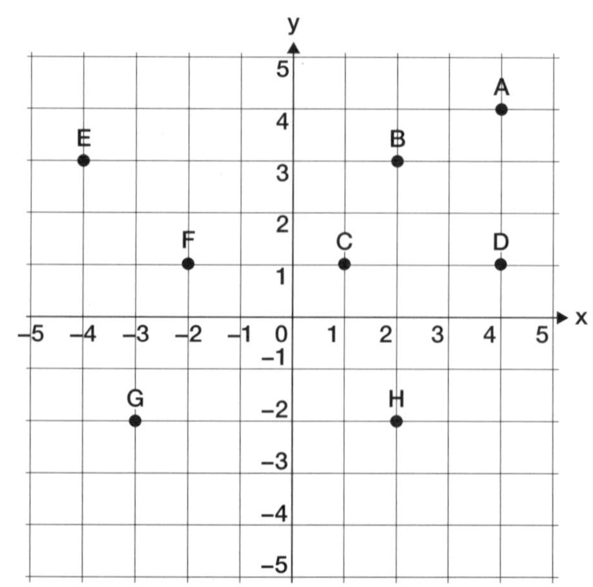

Number and Algebra

SET 3 Fractions of a quantity

Find the given fraction of each number.

1 $\frac{1}{4}$ of 20
2 $\frac{1}{5}$ of 30
3 $\frac{1}{2}$ of 12
4 $\frac{1}{3}$ of 9
5 $\frac{2}{5}$ of 15
6 $\frac{2}{3}$ of 21
7 $\frac{4}{5}$ of 20
8 $\frac{3}{10}$ of 40
9 $\frac{5}{8}$ of 40
10 $\frac{3}{4}$ of 40

True or false?

11 $\frac{1}{2}$ of 30 $<$ $\frac{1}{4}$ of 20
12 $\frac{1}{5}$ of 30 $>$ $\frac{1}{3}$ of 24
13 $\frac{1}{3}$ of 18 $<$ $\frac{2}{3}$ of 12
14 $\frac{2}{3}$ of 24 $>$ $\frac{3}{4}$ of 20
15 $\frac{3}{4}$ of 20 $=$ $\frac{1}{2}$ of 40

16

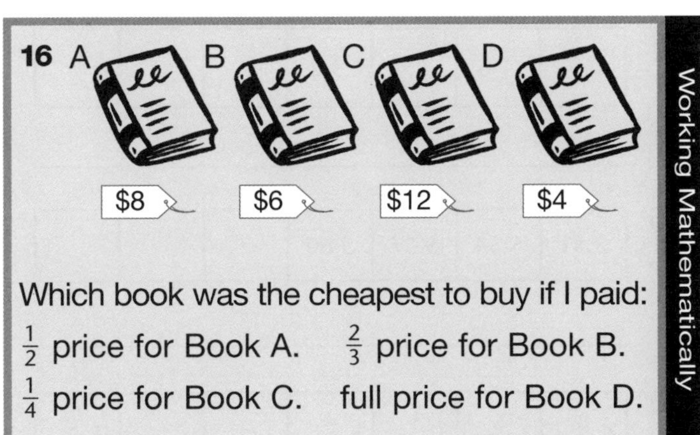

Which book was the cheapest to buy if I paid:
$\frac{1}{2}$ price for Book A. $\frac{2}{3}$ price for Book B.
$\frac{1}{4}$ price for Book C. full price for Book D.

SET 4 Extension

1 $\frac{3}{8}$ of 96 = $\frac{3}{6}$ of 72. True or false?
2 Round 35 244 to the nearest 10.
3 How many 120 g masses would be needed to balance a 2.4 kg mass?
4 Average of 3.5, 2.5, 3.2 and 2.8
5 Round 25 369 to the nearest 1000.
6 What is the value of 6 in 6937.24?
7 Is body temperature about 5°C, 0°C, 100°C or 37°C?
8 Write five million, three hundred and two thousand in figures.
9 Area of a square with a 15 cm base
10 Estimate the answer for 39 987 × 3.
11 How much is 4.6 kg at $9 per 500 g?
12 Convert 1655 m into kilometres.
13 4.35 tonnes = ☐ kilograms
14 How many seconds in 1.3 hours?
15 Round 450.97 to the nearest 100.
16 $\frac{1}{4}$ of 28 × $\frac{6}{10}$ of 130
17 How many sides are on 7 decagons?
18 What is the perimeter of a decagon with 17.2 cm sides?

Measurement Mass units

Convert the kilograms to the nearest whole tonne.

1 Take-off weight 394 632 kg ☐ t
2 Landing weight 295 747 kg ☐ t
3 Zero fuel weight 244 944 kg ☐ t
4 What is the difference between the take-off weight and the landing weight?

Complete the chart showing the difference between gross mass and net mass.

	Gross mass	Net mass	Difference
5	2 kg	1950 g	
6	1 kg		60 g
7		4810 g	190 g
8	5 kg	4550 g	
9	4.5 kg	3995 g	
10	2.25 kg	1800 g	
11		2990 g	260 g

Number and Algebra

SET 1 Basic

1 31 ÷ 6

2 2000 mL = ☐ L

3 34 × 100

4 9 ☐ 3 = 27

5 70 ☐ 80 = 150

6 8 × 8 + 6

7 60 – 45

8 Divide 15 by 4.

9 How many 100s in 3754?

10 What is the value of 5 in 8754?

11 $1\frac{1}{2}$ kg = ☐ g

12 I had $50 but spent $27. How much have I left?

13 37 426 + 300

14 How much in $1\frac{1}{2}$ kg of meat at $8 per kg?

15 How much are 5 pots at $9 each?

16 How many prime numbers are there between 6 and 20?

☐ prime numbers

SET 2 Decimal/fraction number patterns

Complete the sequence, then write a rule for each one.

1

$\frac{3}{10}$	$\frac{6}{10}$	$\frac{9}{10}$	$1\frac{2}{10}$			

____

2

$4\frac{7}{10}$	$4\frac{3}{10}$	$3\frac{3}{10}$	$3\frac{5}{10}$			

____

3

1.7	2	2.3	2.6			

____

4

0.8	1.6	2.4	3.2			

____

5

5.51	5.54	5.57	5.60			

____

6

$2\frac{8}{8}$	$2\frac{6}{8}$	$3\frac{1}{8}$	$3\frac{4}{8}$			

____

Statistics and Probability Median score

Find the median score in these sets of scores.

Find the median in this **odd** number of scores.	Find the median in this **even** number of scores. You will need to average the 2 median scores.
1 14, 15, 26, 27, 39, 40, 41, 42, 45 ______	6 33, 34, 39, 41, 43, 55, 61, 70 ______
2 33, 34, 45, 47, 58, 59, 60, 61, 66 ______	7 71, 93, 94, 105, 115, 125, 130, 136 ______
3 24, 26, 48, 50, 74, 76, 78, 80, 90 ______	8 66, 69, 75, 82, 100, 188, 195, 220 ______
4 55, 57, 62, 70, 80, 91, 97, 99, 105 ______	9 222, 242, 262, 285, 315, 322, 342, 362 ______
5 113, 114, 116, 118, 208, 212, 219 ______	

Number and Algebra

SET 3 Order of operations

Substitute the answers in the boxes into the number sentences to see which number is the missing number. Circle your choice.

		Answers		
1	(5 + 4) × ☐ = 54	5	6	7
2	6 + (3 × ☐) = 21	3	4	5
3	(☐ ÷ 6) + 8 = 11	18	19	20
4	54 ÷ 3 ÷ ☐ = 9	2	8	9
5	(42 ÷ 7) ÷ ☐ = 3	6	3	2
6	☐ × 4 ÷ 5 = 20	20	25	30
7	(19 – ☐) ÷ 7 = 2	9	7	5
8	36 – 6 × ☐ = 12	5	9	4

9 Below the problem are two number sentences. Tick the one you think solves the problem.

Problem: David earns $500 per week and pays $125 tax out of that money. How much 'take home pay' does he earn per year?

($500 – $125) × 52 = $19 500 ☐

$500 × 52 – $125 = $25 875 ☐

Working Mathematically

SET 4 Extension

1 8.75 tonnes = ☐ kg

2 How much is 14.25 kg of potatoes at $3 per kilogram?

3 Average 125, 275, 100 and 120

4 All quadrilaterals are polygons. True or false?

5 How many combinations can I make with 4 shirts and 3 hats?

6 What percentage of $120 is $24?

7

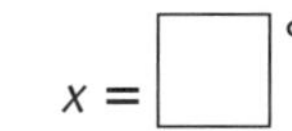

$x =$ ☐°

8 John travelled for 6 hours and covered 522 km. What was his average speed?

9 18 m, less 180 cm

10 (17.86 + 2.14) × 6

11 3.26 × 100

12 Order these decimals: 0.1, 1.0, 0.11, 11.1.

13 How much money did I begin with if I spent 75% of my money and had $50 left over?

____

Working Mathematically

Measurement Square and cubic metres

1 Calculate the area of the floor of the workshop. ☐ m^2

2 Calculate the volume of the workshop. ☐ m^3

3 Calculate the cost of building the workshop at $300 per m^3. $ ☐

UNIT 34

Number and Algebra

SET 1 Basic

1 350 + 150

2 500 ☐ 100 = 5

3 360 – 90

4 40 ☐ 170 = 210

5 Product of 10 and 60

6 How many days in June, July and August?

7 $\frac{3}{100}$ = 0.3. True or false?

8 76 × 1000

9 Tenths in $2\frac{1}{2}$

10 $9^2 - 3$

11 $30 - 5 = 5^2$. True or false?

12 Double 280.

13 Triple 20.

14 36 – 7 = (6 × 5). True or false?

15 If 4 balls cost $2.40, how much would 3 balls cost?

SET 2 Number patterns

1 Complete the grid to show the price of pies.

1	2	3	4	5	6	7
$3	$6	$9	$	$	$	$

2 Complete the grid that shows how much milk Sam uses each week.

Sun	Mon	Tues	Wed	Thur	Fri	Sat
1.5 L	3 L	4.5 L				

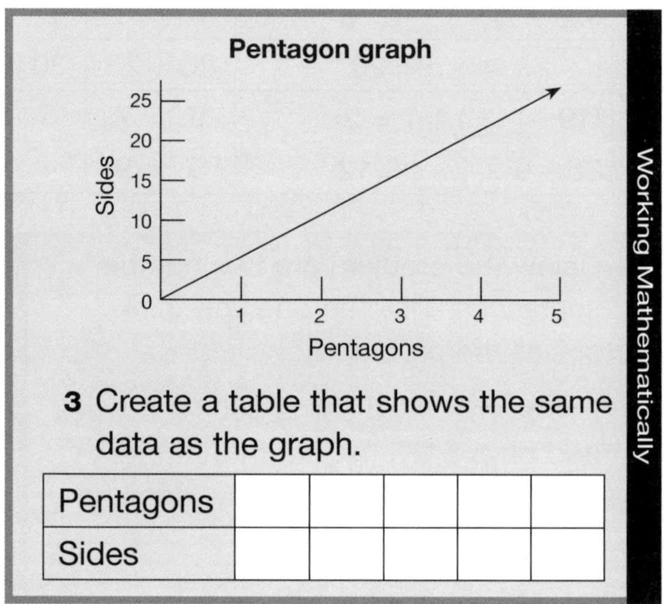

Working Mathematically

3 Create a table that shows the same data as the graph.

Pentagons					
Sides					

Measurement and Space Creating angles

Trace over the two arms of the intersecting lines in order to make a right angle.	Trace over the two arms of the intersecting lines in order to make an angle of 60°.
1 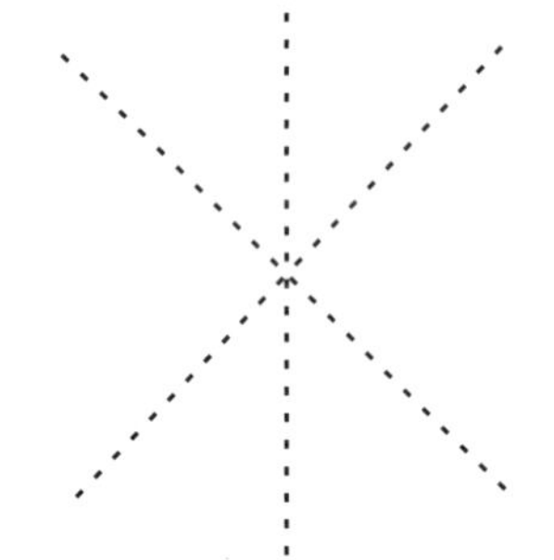	2 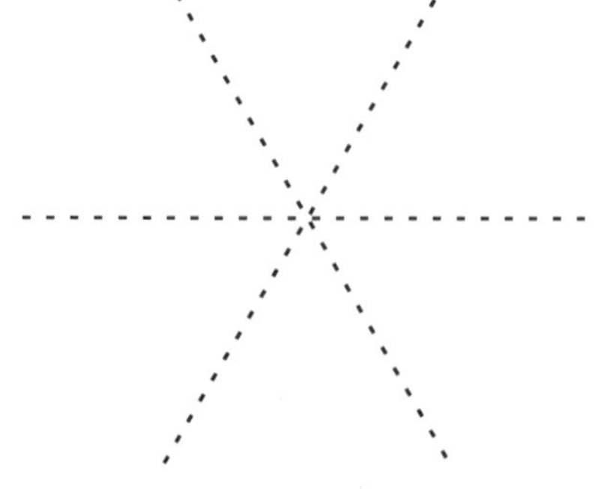
Use different colour pens to trace over any other rectangles that you can see. How many did you see? ______	Adjacent to the 60° angle you identified, trace over two more arms to create an angle of 120°. What is the sum of the two angles you have created? ______

Number and Algebra

SET 3 Finding whole quantities

1 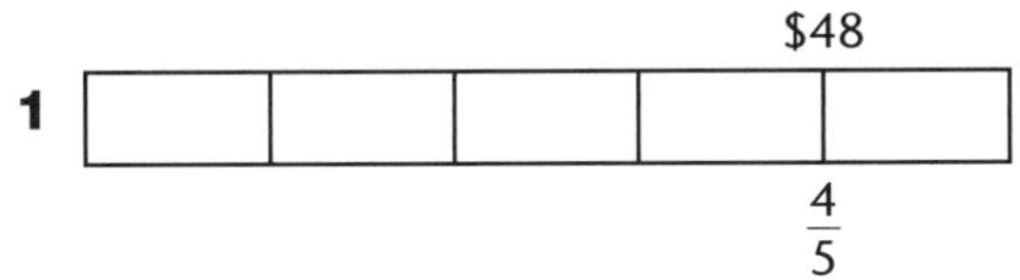

If $\frac{4}{5}$ of an amount is $45, what is the total amount? ______________

2 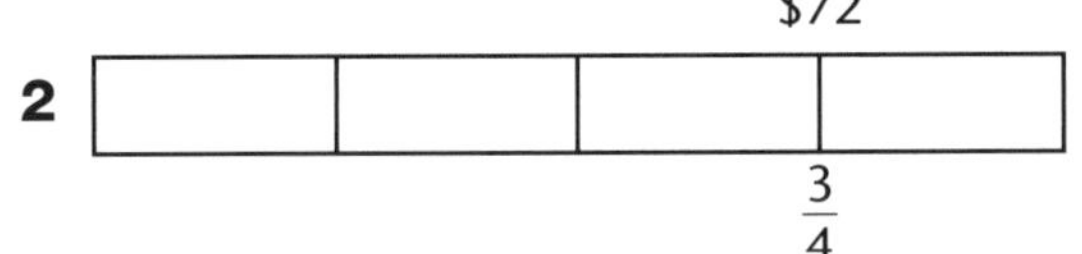

If $\frac{3}{4}$ of an amount is $72, what is the total amount? ______________

3 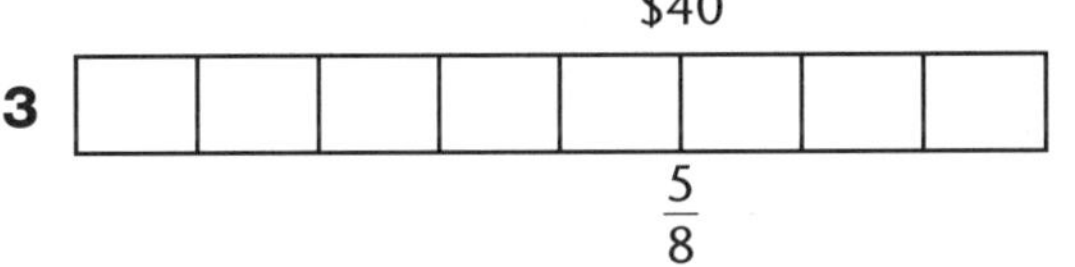

If $\frac{5}{8}$ of an amount is $40, what is the total amount? ______________

4 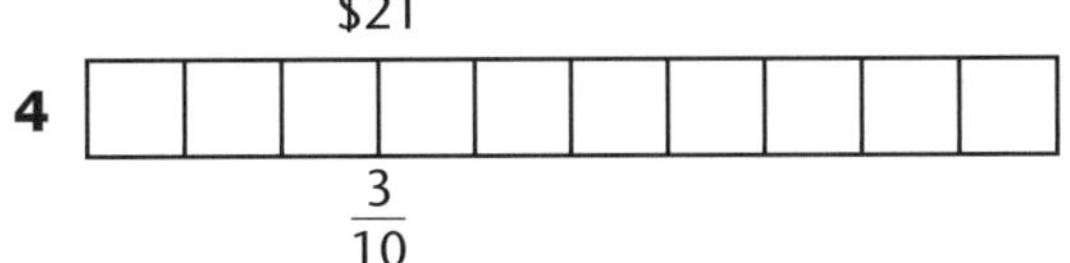

If $\frac{3}{10}$ of an amount is $21, what is the total amount? ______________

5 Sanya has run 15 km in the half marathon, which is equal to 5/6 of the course. How long is the full course?

____

SET 4 Extension

1 $250 \times 3 + 4$

2 $\frac{7}{10}$ of 50 – 19.5

3 Average of 27, 92, 53, 20

4 Emil travelled 470 km in 5 hours. What was his average speed?

5 Which one is not equivalent: $\frac{3}{5}$, 33% or 0.6?

6 $\frac{5}{8} \times \$72 - \frac{3}{4}$ of 60

7 How much are 12 toys at $10.50 each?

8 Which is larger: $\frac{3}{4}$ or $\frac{3}{5}$?

9 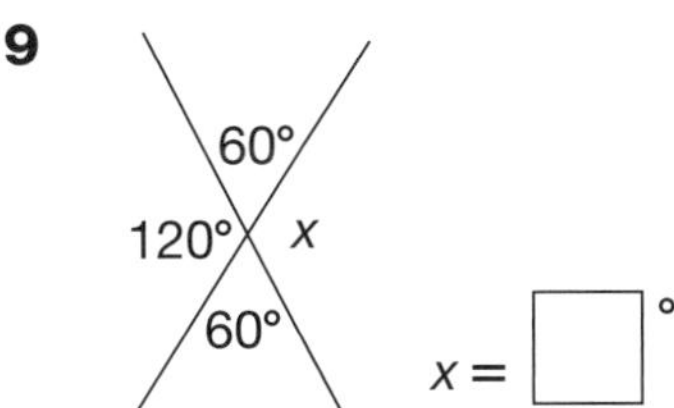

10 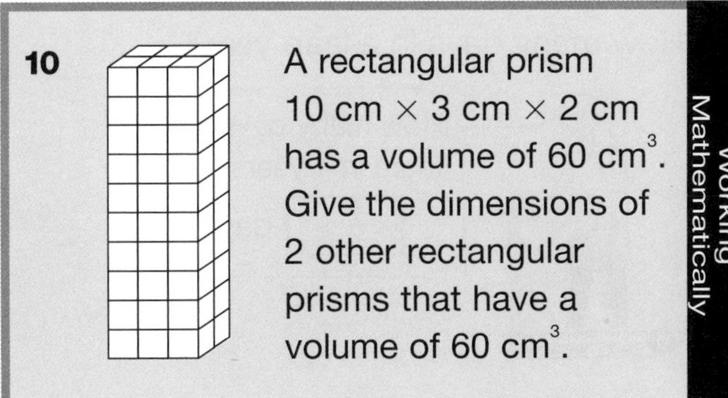

A rectangular prism 10 cm × 3 cm × 2 cm has a volume of 60 cm³. Give the dimensions of 2 other rectangular prisms that have a volume of 60 cm³.

Working Mathematically

Statistics and Probability Stacked column graphs

Calculate the differences in temperature.

1 Minimum temperature of Auckland and Paris.

2 Maximum temperature of Moscow and London.

3 Maximum and minimum temperatures of Rome.

4 Minimum London and maximum Moscow.

5 Minimum Auckland and minimum Moscow.

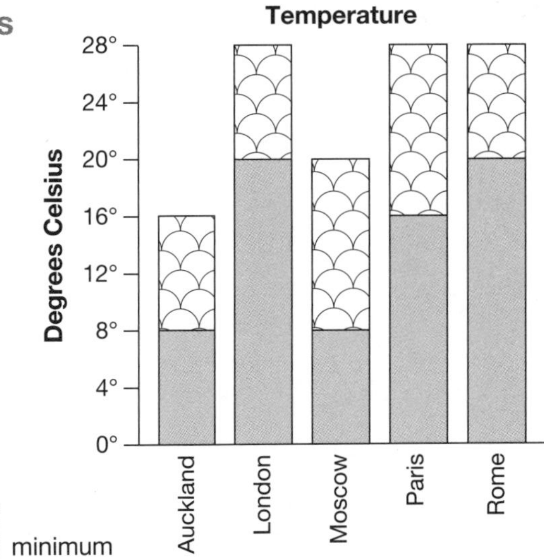

Key
maximum
minimum

UNIT 35

Number and Algebra

SET 1 Basic

1 $49 + 9$

2 $200 ÷ 10

3 $100 \div 5$

4 8 ☐ 5 = 1 r 3

5 $65 + 35$

6 126 ☐ 24 = 102

7 $250 - 36$

8 $9 \times 7 + 3^2$

9 500 mL = ☐ L

10 Is 33 a prime number?

11 11:57 + 10 minutes

12 Value of 6 in 63 237

13 3 thousands + 36 241

14 Hours in 3 days

15 How many days in a leap year?

16 

How many days are there in 4 years? ☐ days

SET 2 Equivalent number sentences

Supply the missing numbers.

1 $5 \times \square = 40$

2 $6 + 3 + \square = 21$

3 $\square \div 8 = 6$

4 $50 - \square - 3 = 20$

5 $20 \times \square = 100$

6 $\square \times 1 = 87$

7 $5^2 + \square = 41$

Supply the missing numbers to create equivalent number sentences.

8 $3 \times 4 = \square + 5$

9 $15 \div \square = 20 \div 4$

10 $7 + \square = 5 \times 20$

11 $\square \times 3 = 9 \times 4$

12 $6^2 + 6 = \square \times 6$

13	2	3	4	6	8

Select numbers from the grid to make these number sentences equivalent.

$48 \div \square = \triangle \times \bigcirc$

Working Mathematically

Space Diagonals

1 Draw the diagonals on the polygons.

2 Draw a line to match each polygon to a label.

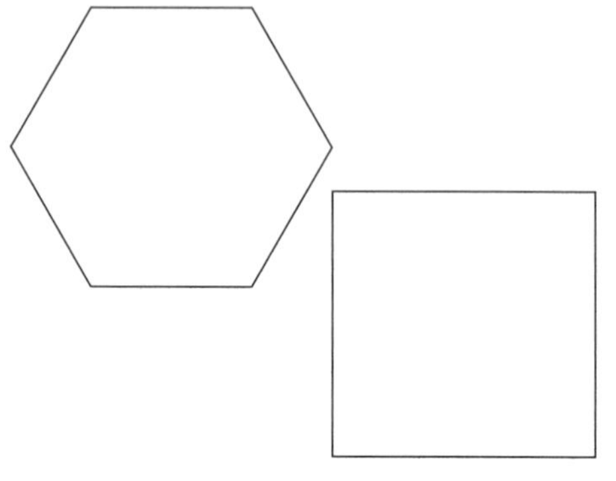

2 diagonals

0 diagonals

5 diagonals

9 diagonals

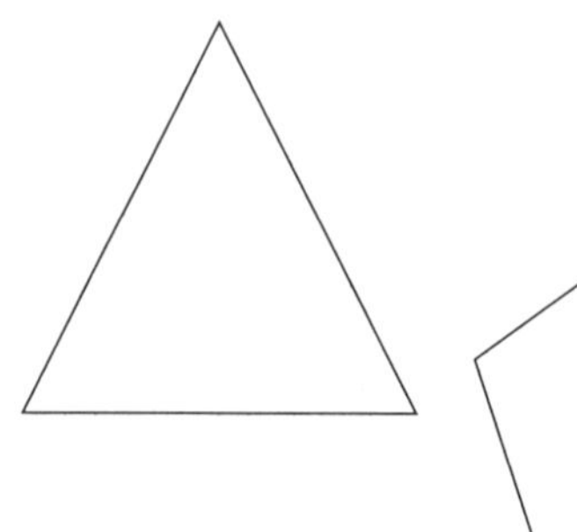

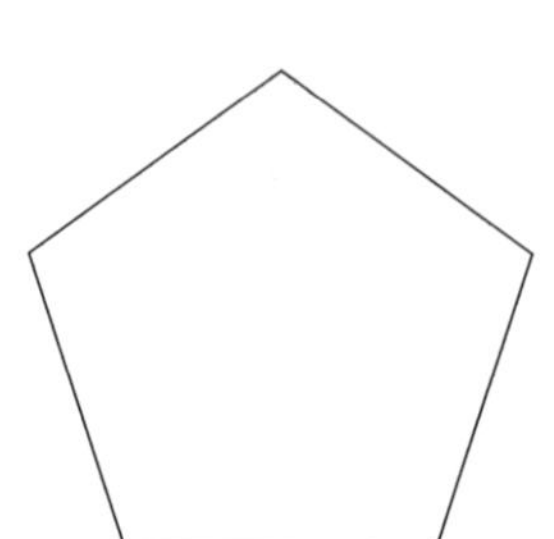

Number and Algebra

SET 3 Negative numbers

Display the answers to these questions on the number lines. A starting point has been given.

1 3 + 4 – 13

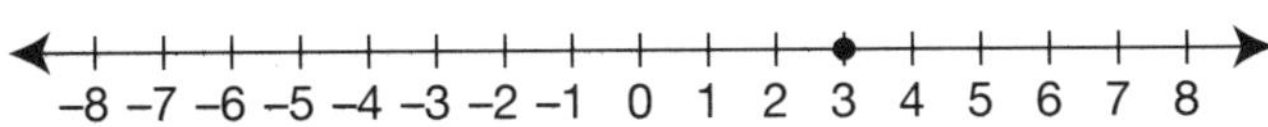

2 1 + 2 – 8

–8 –7 –6 –5 –4 –3 –2 –1 0 1 2 3 4 5 6 7 8

3 8 – 4 – 12

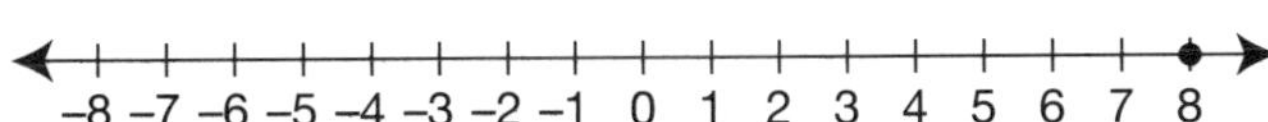

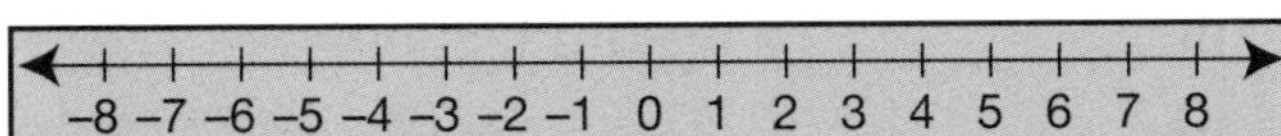

Use the number line to answer the questions.

4 8 – 5 – 3 =

5 4 – 6 – 1 =

6 5 – 8 + 1 =

7 3 – 7 + 3 =

8 8 – 8 – 8 =

9 2 – 4 + 5 =

10 3 – 2 + 7 =

11 4 – 5 – 6 =

12 0 – 6 + 7 =

13 8 – 6 – 5 + 3 =

14 6 – 8 – 5 + 7 =

15 –3 + 9 – 7 =

SET 4 Extension

1 25 × 20 = 500. True or false?

2 Perimeter of a regular nonagon with sides of 24 cm

3 Average of 3.75, 4.25, 5.9 and 2.1

4 How many kilograms in 14.75 tonnes?

5 3.5 × 10

6 Round and estimate 119.8 × 6.9.

7 How many 125 g packets of peanuts in 4 kg?

8 Round off 2499 then divide by 5.

9

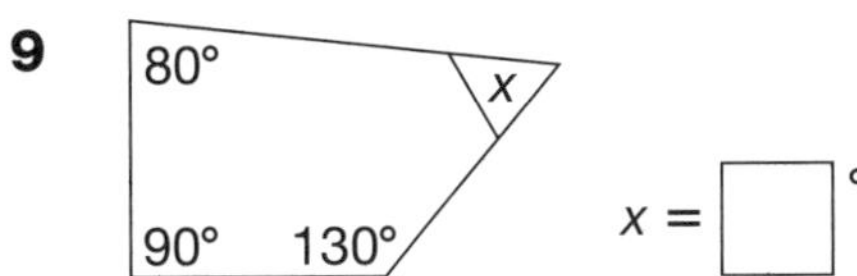

10 Parallel lines never meet. True or false?

11 Write 11:27 pm in 24-hour time.

12 3.5 hours at an average speed of 80 km per hour = ☐ km

13 Minutes in 1.5 days

14 Which does not fit: 0.6, 60%, $\frac{6}{100}$ or $\frac{6}{10}$?

15 Apply 6 × (■ + 3) – 7 = ●

■	3	1	5	20	40	72
●						

Statistics and Probability Data exploration

Recently 50 tickets were sold to the Youth Club Freak Night. Use the illustration showing the age of each ticket purchaser to construct a tally and frequency table to represent the ages of the people attending. It has been started for you.

14 14 14 15 15 15 15
15 16 16 16 16 16 16
16 17 17 17 17 17 17
17 17 17 17 17 17 17
18 18 18 18 18 18 18
18 18 18 18 18 18 18
18 19 19 19 19 19 19 19

Age	Tally	Frequency
14	|||	3
15		
16		
17		
18		
19		

Maths helpers

Length

10 millimetres (mm) = 1 centimetre (cm)

100 centimetres (cm) = 1 metre (m)

1000 metres (m) = 1 kilometre (km)

Mass

1000 grams (g) = 1 kilogram (kg)

1000 kilograms (kg) = 1 tonne (t)

Capacity

1000 millilitres (mL) = 1 litre (L)

Time

60 seconds = 1 minute

60 minutes = 1 hour

24 hours = 1 day

7 days = 1 week

14 days = 1 fortnight

12 months = 1 year

52 weeks = 1 year

365 days = 1 year

366 days = 1 leap year

10 years = 1 decade

100 years = 1 century

Months of the year

Thirty days has September, April, June and November. All the rest have thirty-one, except February alone, which has twenty-eight days clear and twenty-nine days each leap year.

Seasons

Summer: December, January, February

Autumn: March, April, May

Winter: June, July, August

Spring: September, October, November

Roman numerals

1 = I

2 = II

3 = III

4 = IV

5 = V

6 = VI

7 = VII

8 = VIII

9 = IX

10 = X

20 = XX

30 = XXX

40 = XL

50 = L

60 = LX

70 = LXX

80 = LXXX

90 = XC

100 = C

500 = D

1000 = M

Multiplication facts

×	0	1	2	3	4	5	6	7	8	9	10
0	0	0	0	0	0	0	0	0	0	0	0
1	0	1	2	3	4	5	6	7	8	9	10
2	0	2	4	6	8	10	12	14	16	18	20
3	0	3	6	9	12	15	18	21	24	27	30
4	0	4	8	12	16	20	24	28	32	36	40
5	0	5	10	15	20	25	30	35	40	45	50
6	0	6	12	18	24	30	36	42	48	54	60
7	0	7	14	21	28	35	42	49	56	63	70
8	0	8	16	24	32	40	48	56	64	72	80
9	0	9	18	27	36	45	54	63	72	81	90
10	0	10	20	30	40	50	60	70	80	90	100

Addition facts

+	2	3	4	5	6	7	8	9	10	11	12
2	4	5	6	7	8	9	10	11	12	13	14
3	5	6	7	8	9	10	11	12	13	14	15
4	6	7	8	9	10	11	12	13	14	15	16
5	7	8	9	10	11	12	13	14	15	16	17
6	8	9	10	11	12	13	14	15	16	17	18
7	9	10	11	12	13	14	15	16	17	18	19
8	10	11	12	13	14	15	16	17	18	19	20
9	11	12	13	14	15	16	17	18	19	20	21
10	12	13	14	15	16	17	18	19	20	21	22
11	13	14	15	16	17	18	19	20	21	22	23
12	14	15	16	17	18	19	20	21	22	23	24

UNIT 1 Number and Algebra

SET 1

1 32
2 48
3 17
4 ×
5 ÷
6 81
7 9
8 ×
9 55
10 24
11 9
12 869
13 72
14 1, 3, 5, 9, 15, 45
15 $6.50

SET 2

1 5908
2 8347
3 4542
4 11 809
5 5332
6 11 983
7 $3 999
8 $14 729

SET 3

1 15, 25, 35, 40, 30
2 14, 35, 49, 63, 56
3 18, 45, 63, 81, 72
4 210
5 320
6 240
7 420
8 400
9 90
10 52
11 104
12 168
13 72
14 104
15 168
16 256
17 120

SET 4

1 9.4
2 800
3 8000
4 20
5 8
6 6.1
7 $8.25
8 110
9 90
10 2000
11 True
12 13°
13 4 km
14 About 8000
15 $1.35
16 15

Space

1 45°
2 150°
3 60°
4 120°

Measurement

1 50 km
2 10 km
3 30 km
4 25 km
5 35 km
6 45 km

UNIT 2 Number and Algebra

SET 1

1 11
2 0
3 7
4 250
5 49
6 12 February
7 7
8 $5
9 71
10 Winter
11 ÷
12 4
13 20
14 7000
15 75

SET 2

1 86
2 143
3 148
4 135
5 177
6 464
7 867
8 1156
9 47
10 32
11 34
12 112
13 201
14 614

	Question	Rounded to 100	Approximate answer
15	395 + 206	400 + 200	600
16	591 – 298	600 – 300	300
17	513 + 387	500 + 400	900
18	785 – 589	800 – 600	200
19	372 + 329	400 + 300	700
20	882 – 286	900 – 300	600

SET 3

1 $\frac{10}{100}$, 0.10, 10%
2 $\frac{25}{100}$, 0.25, 25%
3 $\frac{7}{10}$, 0.7, 70%
4 $\frac{20}{100}$, 0.20, 20%
5 $\frac{1}{2}$, 0.5, 50%
6 $\frac{1}{4}$, 0.25, 25%
7 $\frac{3}{4}$, 0.75, 75%
8 $\frac{27}{100}$, 0.29, 30%
9 0.33, 35%, $\frac{53}{100}$
10 9%, 0.9, $\frac{99}{100}$
11 49%, $\frac{1}{2}$, 0.54
12 4%, 0.21, $\frac{3}{10}$
13 $\frac{9}{100}$, 90%, 0.95
14 0.03, 7%, $\frac{70}{100}$

SET 4

1 90
2 23%
3 6 ones
4 35
5 $72
6 6
7 63 200
8 $31.00
9 True
10 803
11 9 × 6 = 54
12 3 × 2 = 6
13 9 ÷ 3 = 3
14 9 × 3 + 6 = 33
15 72 ÷ 9 × 6 = 48
16 54 ÷ 6 = 3 × 3

Space

1 2 3 4

Measurement

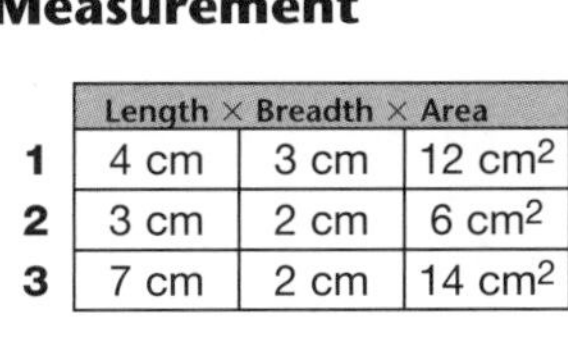

	Length ×	Breadth ×	Area
1	4 cm	3 cm	12 cm^2
2	3 cm	2 cm	6 cm^2
3	7 cm	2 cm	14 cm^2

UNIT 3 Number and Algebra

SET 1

1 15
2 12
3 9
4 ×
5 –
6 9
7 6
8 ÷
9 28
10 21
11 7
12 70
13 Yes
14 1216
15 250

SET 2

1 2000
2 4000
3 2425
4 4960
5 3561
6 5111
7 5172
8 4182
9 2142
10 5086
11 4732
12 3746

SET 3

1 323
2 142
3 35
4 26
5 109
6 93
7 $8 each
8 18
9 45
10 51
11 106 each
12 $132
13 54
14 31

SET 4

1 70 000
2 8%
3 180
4 4
5 43
6 3158
7 76
8 35°C
9 7300
10 29 000
11 60
12 7.5 cm
13 4
14 $\frac{24}{40}$
15 0.15, 15% or 15 hundredths
16 $31.50

Statistics and Probability

1 12
2 28
3 22
4 14
5 12

Space

Possible solutions:

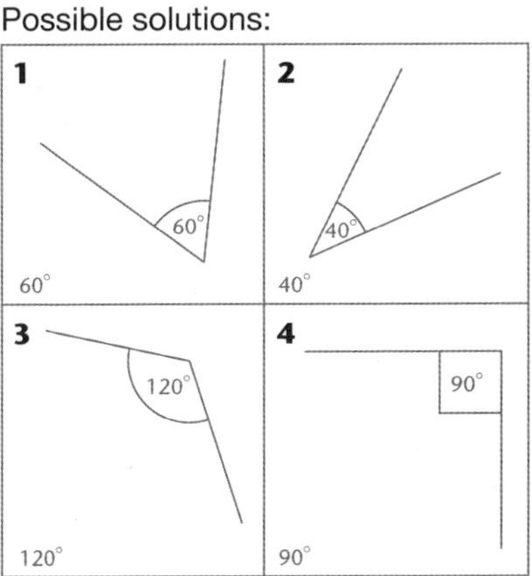

Answers

UNIT 4 Number and Algebra

SET 1

1 45
2 7
3 22
4 16
5 –
6 ÷
7 5
8 +
9 $7
10 5
11 1, 3, 9
12 Yes
13 No
14 700
15 $11

SET 2

1 4290 m – $8.58
2 5335 m – $10.67
3 5420 m – $10.84
4 5060 m – $10.12
5 8300 m – $16.60
6 6735 m – $13.47
7 7950 m – $15.90

SET 3

1 Hun Thou Ten Thou Thou Hund Tens Ones
2 Hun Thou Ten Thou Thou Hund Tens Ones
3 64 928
4 52 818, 52 819
5 193
6 47 000
7 157 695
8 252 340
9 349
10 443 186
11 214 900
214 899
12 25 133
13 975 970

SET 4

1 75
2 9
3 700
4 5704, 5804
5 6
6 243
7 $19
8 140
9 3
10 $6.80
11 250
12 $\frac{1}{4}$, $\frac{1}{3}$, $\frac{1}{2}$, $\frac{3}{4}$
13 10:23
14 $6.48
15 $14.45
16 $6.50
17 About 25 000
(50 × 500)

Number and Algebra

1 –1
2 –7
3 –6

Measurement and Space

	Name	Faces	Vertices	Edges
1	Cylinder	3	0	2
2	Rectangular pyramid	5	5	8
3	Rectangular prism	6	8	12
4	Triangular prism	5	6	9

UNIT 5 Number and Algebra

SET 1

1 24
2 36
3 9
4 ×
5 ÷
6 64
7 8
8 ×
9 60
10 23
11 6
12 961
13 90
14 1, 3, 9, 27
15 216, 432

SET 2

1 35
2 350
3 3500
4 980
5 9800
6 98 000
7 About 2000
8 4221
9 1251
10 $66.15
11 1052
12 1767
13 2442
14 2616
15 2898
16 2366

SET 3

1 29%
0.29
2 37%
0.37
3 0.1
4 0.5
5 0.05
6 0.75
7 0.5
8 0.27
9 0.09
10 0.97
11 70%
12 25%
13 99%
14 5%
15 35%
16 7%
17 9%
18 75%
19 0.09, 0.35, $\frac{38}{100}$, 70%

SET 4

1 348
2 $1\frac{3}{4}$
3 6000
4 6894, 6884
5 100°C
6 $196
7 40%, 0.47, $\frac{1}{2}$, $\frac{3}{5}$
8 360
9 $46.72
10 306 km
11 95°
12 10 000
13 90
14 10
15 $48.75
16 920 mm

Statistics and Probability

1 Approx 15
2 Approx 10
3 Approx 5

Measurement

1 Coffee 20
Tea 32
Margarine 8
Chocolate 40
Tomato paste 16
Peanut butter 10

2 Margarine
Tomato paste
Tea
Peanut butter

UNIT 6 Number and Algebra

SET 1

1 –
2 24
3 7
4 6
5 160
6 +
7 ÷
8 63
9 18
10 6000
11 1, 2, 4, 5, 10, 20
12 1423
13 No
14 210
15 10

SET 2

1 $1050
2 367 km
3 693
4 327
5 $588
6 $625

SET 3

1 4, 8, 12, 16, 20, 24, 28, 32
2 5, 7, 9, 11, 13, 15, 17, 19
3 11, 22, 33, 44, 55, 66, 77, 88
4 4, 5, 6, 7, 8, 9, 10, 11
Divide by 3
5 6

SET 4

1 105
2 $7.08
3 400 000
4 37
5 0.2, $\frac{1}{4}$, 0.31, 35%
6 $28
7 $19
8 500 m
9 20
10 285
11 $42.82
12 700 m
13 $\frac{36}{100}$
14 $\frac{15}{8}$ or $1\frac{7}{8}$
15 9.3
16 65
17 7.481 5.373 2.299
18 7.434 5.326 2.252
19 7.931 5.823 2.749
20 7.432 5.324 2.250

Statistics and Probability

1 Summer
2 Summer
3 Spring
4 $2250

Measurement

1 2:30 am
2 7:15 pm
3 4:45 pm
4 7:25 pm
5 10:23 pm

UNIT 7 Number and Algebra

SET 1

1 8
2 35
3 16
4 49
5 18
6 460
7 42
8 $3
9 ÷
10 9
11 Autumn
12 No
13 9
14 32
15 192

SET 2

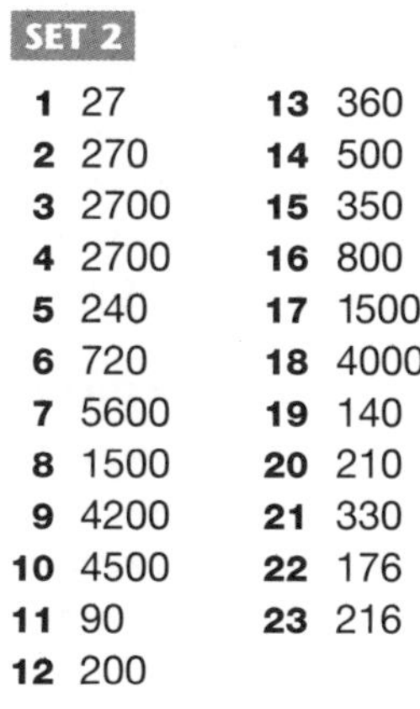

1 27
2 270
3 2700
4 2700
5 240
6 720
7 5600
8 1500
9 4200
10 4500
11 90
12 200
13 360
14 500
15 350
16 800
17 1500
18 4000
19 140
20 210
21 330
22 176
23 216

SET 3

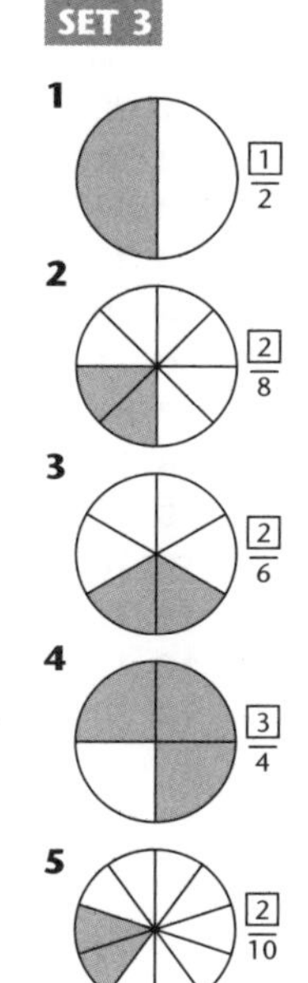

1 $\frac{1}{2}$
2 $\frac{2}{8}$
3 $\frac{2}{6}$
4 $\frac{3}{4}$
5 $\frac{2}{10}$
6 True
7 False
8 True

SET 4

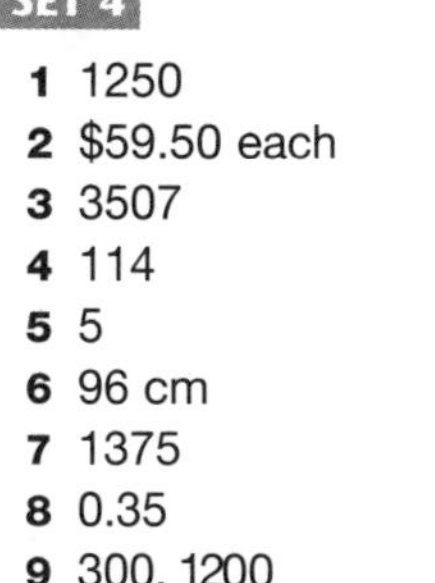

1 1250
2 $59.50 each
3 3507
4 114
5 5
6 96 cm
7 1375
8 0.35
9 300, 1200
10 500, 4000
11 300, 1800
12 900, 2700
13 500, 2500
14 900, 8100

Statistics and Probability

20 marbles

Red $\frac{4}{10}$

Blue $\frac{2}{10}$

Green $\frac{1}{10}$

Gold $\frac{3}{10}$

Measurement

1 100 mm
2 75 mm
3 110 mm
4 102 mm
5 108 mm
6 30 mm
7 70 mm
8 180 mm
9 95 mm
10 115 mm

UNIT 8 Number and Algebra

SET 1

1 8
2 35
3 24
4 27
5 +
6 ÷
7 30
8 3
9 ×
10 400
11 $\frac{7}{10}$
12 1, 3, 5, 15
13 Yes
14 15
15 1

SET 2

1 53 334
2 34 521
3 29 029
4 55 691
5 95 952

SET 3

1 9
2 1.2
3 6
4 1.5
5 5
6 3
7 13
8 0.5
9 42
10 8
11 ×, +
12 +, ×
13 −, ×
14 ×, −

SET 4

1 $13\frac{1}{4}$
2 40
3 3100
4 $\frac{3}{4}$
5 15°C
6 1 226 000
7 24
8 4
9 60°
10 3.5, 3.75, $3\frac{9}{10}$, $4\frac{1}{4}$
11 $32
12 6
13 33

Measurement and Space

1 170 mm
2 200 mm
3 120 mm
4 100 mm
5 100 mm
6 120 mm

Statistics and Probability

1 30
2 25
3 5
4 10
5 25

UNIT 9 Number and Algebra

SET 1

1 49
2 ×
3 ÷
4 64
5 28
6 45
7 7
8 63
9 38
10 3752
11 No
12 165
13 30%
14 $3.30
15 Saturday 14 July

SET 2

1 2043
2 369
3 1123
4 658
5 387
6 651
7 $1376\frac{3}{5}$
8 $1825\frac{2}{4}$
9 $1952\frac{1}{3}$
10 $1232\frac{2}{6}$
11 $825\frac{3}{8}$
12 $1452\frac{4}{5}$
13 364 with 1 left over
14 145 with 3 left over

SET 3

1 True
2 True
3 True
4 True
5 False
6 True
7 True
8 False
9 True
10 True
11 $1\frac{1}{3}$
12 $1\frac{2}{5}$
13 $2\frac{1}{4}$
14 $1\frac{2}{6}$
15 $2\frac{1}{5}$
16 $2\frac{1}{2}$

Possible solutions:

17 $\frac{5}{2}$, $\frac{10}{4}$
18 $\frac{5}{4}$, $\frac{10}{8}$
19 $\frac{4}{3}$, $\frac{8}{6}$
20 $\frac{13}{6}$, $\frac{26}{12}$

SET 4

1 370
2 31 004
3 7
4 125
5 20 000
6 105
7 33
8 0.51
9 2350
10 290 006
11 $2
12 $34

Space

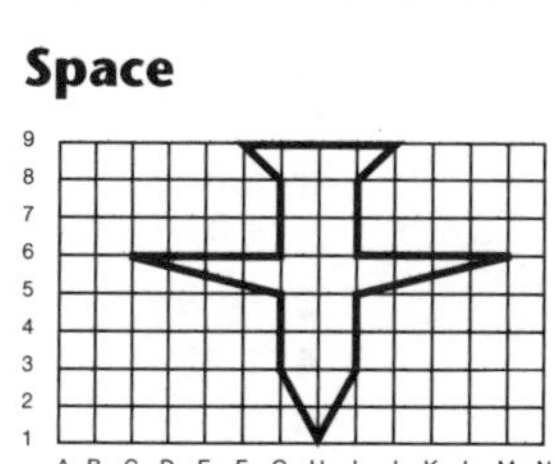

Statistics and Probability

1 (Y)(R)(Y)(R) (R)(Y)(Y)(R)
(R)(R)(Y)(Y) (Y)(R)(R)(Y)
(R)(Y)(R)(Y) (Y)(Y)(R)(R)

2 7 is the more likely score. There are 6 combinations of scores, that have a sum of 7 but only 3 that have a sum of 10.

Answers

UNIT 10 Number and Algebra

SET 1

1 16
2 56
3 15
4 +
5 ×
6 25
7 7
8 –
9 42
10 9
11 1324
12 27
13 No
14 3175
15 4
16 12

SET 2

1 4, 7, 10, 13, 16, 19, 22, 25
2 3, 6, 9, 12, 15, 18, 21, 24
3 15, 21, 28, 36, 45

SET 3

1 Prime
2 Composite
3 Composite
4 Composite
5 Prime
6 Composite
7 Composite
8 Prime
9 Prime
10 Composite
11 7
12 17, 19
13 23, 29
14 53, 59
15 97, 101, 103, 107, 109

SET 4

1 90
2 True
3 475
4 332
5 10
6 11.16
7 8
8 150
9 100°C
10 105 226
11 37 190
12 1
13 80%
14 15 992 (95 952 ÷ 6)

Space

1 Scalene triangle
2 Equilateral triangle
3 Isosceles triangle
4 Right-angle triangle

Measurement

1 24 cm^2, 12 cm^2
2 40 cm^2, 20 cm^2
3 12 cm^2, 6 cm^2
4 50 cm^2, 25 cm^2
5 20 cm^2, 10 cm^2

UNIT 11 Number and Algebra

SET 1

1 8 r 1
2 15
3 W
4 39
5 730
6 61
7 20 cm
8 549
9 25
10 0.21
11 0
12 3 r 2
13 61
14 678
15 9
16 15

SET 2

1 18
2 6
3 9
4 3
5 4
6 12

7 30
8 20
9 30
10 40
11 60
12 60
13 30
14 80
15 50
16 160
17 80
18 60

SET 3

1 Ones
2 Tens
3 Hundreds
4 Tenths
5 Hundredths
6 Thousands
7 Hundredths
8 Thousandths
9 7.7
10 3.22
11 6.291
12 7.07

SET 4

1 $\frac{22}{40}$
2 1710
3 60 cm^3
4 135 min
5 0°C
6 320 m^2
7 $19
8 5
9 40°
10 $34.50
11 4350
12 637 430
13 8
14 74

Space

Statistics and Probability

1

IMPOSSIBLE EVEN CERTAIN

0 0.1 0.2 0.3 0.4 0.5 0.6 0.7 0.8 0.9 1

2 Very unlikely

UNIT 12 Number and Algebra

SET 1

1 180
2 120
3 –
4 –
5 ×
6 7 r 1
7 +
8 0
9 147
10 $\frac{9}{10}$
11 60 000
12 No
13 True
14 3270
15 31
16 $13.50

SET 2

1 5.8
2 5.2
3 24.9
4 1.32
5 13.5, 13.6, 13.65, 13.71
6 4.2
7 1.1
8 266.87
9 $69.69
10 $745
11 $142.56
12 $492.60
13 $3810.47
14 $843.48
15 $1180.38

SET 3

1 Colour any 6 parts
2 Colour one part
3 Colour any 2 parts
4 Colour any 5 parts
5 0.58
6 0.45
7 0.05
8 0.72
9 0.69
10 0.99

11 $\frac{5}{10}$ or $\frac{1}{2}$
12 $\frac{5}{100}$
13 $\frac{1}{100}$
14 $\frac{75}{100}$ or $\frac{3}{4}$
15 $\frac{25}{100}$ or $\frac{1}{4}$
16 $\frac{9}{100}$

17 False
18 True

SET 4

1 0.37
2 $\frac{1}{5}$, 30%, 0.31, $\frac{9}{10}$
3 $2\frac{3}{10}$
4 $65
5 7
6 $\frac{5}{8}$
7 5.5
8 10
9 14 cm
10 $3.90
11 $3
12 $21
13 4
14 Sally: 1.63 m Anna: 1.53 m

Statistics and Probability

1 25 kg
2 45 kg
3 6 and 7
4 15 kg
5 20 kg
6 $37\frac{1}{2}$ kg

Measurement

1 120 m^3
2 60 m^3

UNIT 13 Number and Algebra

SET 1

1 +
2 –
3 36
4 9
5 49
6 16
7 39
8 43
9 No
10 Yes
11 2 r 1
12 30 000
13 $1.35
14 30
15 4
16 695

SET 2

1 5.92
2 5.22
3 24.54
4 $28.90
5 $2.55
6 $3
7 8.989 km
8 $54.40
9 $12.14
10 $19.20
11 $4.55
12 1.256
13 $1.55
14 $0.95
15 $1.75
16 $1.95

SET 3

1 9
2 45
3 27
4 8
5 7
6 6
7 9
8 22
9 6
10 13
11 8
12 9
13 48
14 9
15 23

SET 4

1 19.5 m
2 $3.85
3 $145
4 $1\frac{4}{10}$, 1.49, 150%, 1.53
5 125
6 2 soccer fields
7 175
8 24.5 cm^2
9 $\frac{1}{6}$
10 10 000
11 35°
12 $4.10
13 2
14 5750
15 Cube

Space

1 2 cm, 4 cm — 2 cm, 4 cm
2 5 cm — 3 cm, 5 cm
3 3 cm, 3 cm — 3 cm, 3 cm

Measurement

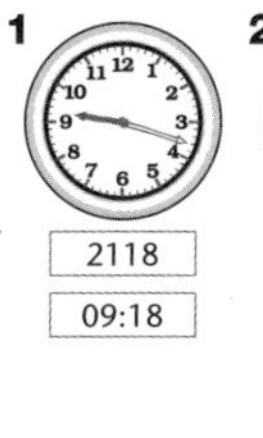

1	2	3	4	5
2118	0405	1713	1344	1142
09:18	04:05	05:13	01:44	11:42

UNIT 14 Number and Algebra

SET 1

1 –
2 21
3 15
4 +
5 6
6 75
7 15
8 13
9 $2.65
10 40
11 Yes
12 8
13 14
14 51 km
15 $48

SET 2

1 489 km
2 3400 t
3 $411
4 18 850 L
5 21 830

SET 3

1
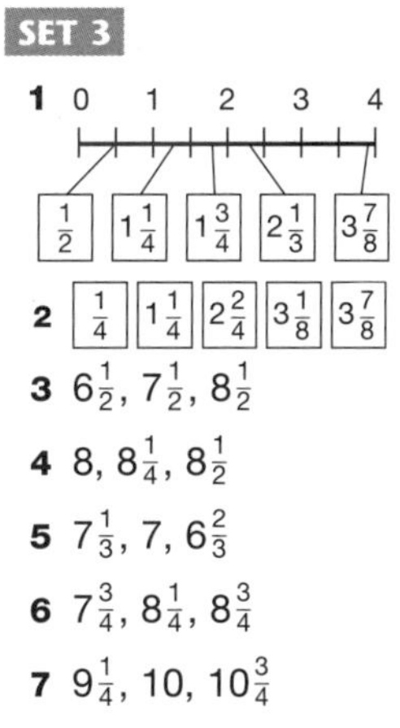

2 $\frac{1}{4}$ $1\frac{1}{4}$ $2\frac{2}{4}$ $3\frac{1}{8}$ $3\frac{7}{8}$
3 $6\frac{1}{2}$, $7\frac{1}{2}$, $8\frac{1}{2}$
4 8, $8\frac{1}{4}$, $8\frac{1}{2}$
5 $7\frac{1}{3}$, 7, $6\frac{2}{3}$
6 $7\frac{3}{4}$, $8\frac{1}{4}$, $8\frac{3}{4}$
7 $9\frac{1}{4}$, 10, $10\frac{3}{4}$
8 $7\frac{2}{5}$, 8, $8\frac{3}{5}$
9 $9\frac{5}{10}$, $10\frac{1}{10}$, $10\frac{7}{10}$

SET 4

1 $180
2 7
3 $16.20
4 $3.00
5 About 12 000
6 1315
7 740
8 $\frac{4}{9}$
9 $\frac{1}{12}$
10 5
11 15
12 8
13 $9.90
14 The park (Pre-school = 400 m Park = 500 m)

Space

	Top view	Front view	Side view
1			
2			

Measurement

Hands on.

UNIT 15 Number and Algebra

SET 1

1 22nd
2 $11\frac{1}{2}$
3 3 r 1
4 ÷
5 90
6 4th
7 2 May
8 7
9 $0.35
10 $0.48
11 20
12 25
13 $6.50
14 64
15 $12.50

SET 2

1 5368
2 6735
3 20 865
4 14 046
5 8432
6 36 281
7 14 872
8 30 245
9 20 196
10 $42 632
11 $82 908

SET 3

1 $\frac{5}{8}$
2 $\frac{8}{10}$
3 $\frac{9}{10}$
4 $\frac{7}{8}$
5 $\frac{12}{8} = 1\frac{4}{8}$
6 $\frac{6}{4} = 1\frac{2}{4}$
7 $\frac{14}{10} = 1\frac{4}{10}$
8 $\frac{7}{5} = 1\frac{2}{5}$
9 $\frac{21}{10} = 2\frac{1}{10}$
10 $\frac{9}{5} = 1\frac{4}{5}$
11 $\frac{6}{10}$
12 $\frac{5}{8}$
13 $\frac{4}{10}$
14 $\frac{1}{5}$
15 $\frac{1}{4}$
16 $\frac{3}{10}$
17 $\frac{1}{8}$
18 $\frac{5}{10}$
19 $\frac{1}{5}$
20 $\frac{6}{10}$

SET 4

1 1 915
2 $5.50 each
3 $8.10
4 116
5 105
6 90 km/h
7 5 250
8 $1.90
9 0.5, 5.0, 15.0, 5^2
10 $146
11 135° or 225° if measured anti-clockwise.
12 The volume increases 8 times.

Statistics and Probability

Discussion: "This year's" column is a lot taller than "last year's", but it only represents $25. Therefore, the title 'large' is not exactly true; 'slight' would be a better word.

Measurement

	Base × height	=	Area
1	4 × 1	=	4 cm^2
2	4 × 2	=	8 cm^2
3	3 × 2	=	6 cm^2
4	3 × 1	=	3 cm^2
5	4 × 2	=	8 cm^2

Answers

UNIT 16 Number and Algebra

SET 1

1 81 **2** 61
3 38 **4** $2.48
5 165 **6** 2025
7 $250 **8** 3.97
9 $5.40 **10** 0.19
11 30 **12** 1483
13 600 **14** 12
15 9 **16** 90

SET 2

1 400 **2** 465
3 1260 **4** 3648
5 2080 **6** 2475
7 10 314 **8** 18 072
9 24 948 **10** 20 382
11 16 994 **12** 40 176

SET 3

1 30 000 + 7000 + 400 + 20 + 3
2 80 000 + 5000 + 600 + 10 + 6
3 20 000 + 5000 + 200 + 9
4 100 000 + 6000 + 10 + 5
5 6 000 000 + 40 000 + 1000 + 500
6 68 499
7 154 347
8 265 358
9 71 358
10 54 812 405
11 480 207
12 4 310 100
13 5 541 040

SET 4

1 $3 × 20 = $60
2 $240
3 $\frac{12}{8} = 1\frac{4}{8} = 1\frac{1}{2}$
4 $9
5 4.88
6 $30
7 75 cm
8 33
9 54
10 37
11 $1\frac{4}{10}$, 1.5, $1\frac{3}{5}$, 1.61
12 $10.10
13 30
14 700 kg
15 $\frac{1}{15}$
16 224 cm³
17 $1\frac{1}{2}$, $1\frac{1}{8}$, $1\frac{2}{5}$, $1\frac{1}{2}$, $1\frac{2}{3}$, $1\frac{1}{2}$, $1\frac{1}{4}$, $2\frac{1}{3}$

Measurement

1

Volume	300 cm³	600 cm³	900 cm³	1000 cm³	1500 cm³	1900 cm³	2000 cm³	2415 cm³	2820 cm³	3000 cm³
Capacity	300 mL	600 mL	900 mL	1 litre	1500 mL	1900 mL	2 litres	2415 mL	2820 mL	3 litres
Mass	300 g	600 g	900 g	1 kilogram	1500 g	1900 g	2 kilograms	2415 g	2820 g	3 kilograms

2 124 g **3** 135 g

Space

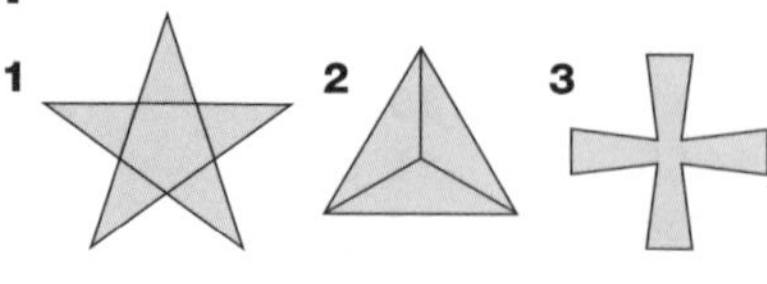

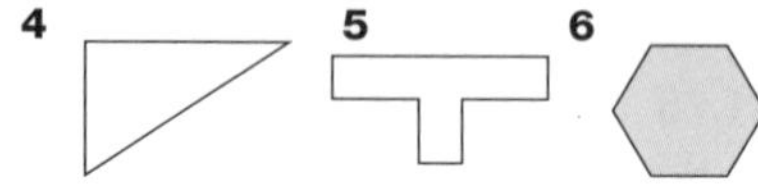

UNIT 17 Number and Algebra

SET 1

1 42
2 63
3 29
4 4
5 ×
6 ÷
7 80
8 8
9 99
10 32
11 52
12 Yes
13 No
14 1887
15 4
16 10

SET 2

1 41
2 161
3 38
4 26
5 105
6 64
7 58
8 188
9 259
10 221
11 55
12 1
13 320
14 $357

SET 3

1 $\frac{2}{4}$
2 $\frac{5}{10}$
3 $\frac{6}{10}$
4 $\frac{2}{8}$
5 $\frac{3}{8}$
6 $\frac{13}{10} = 1\frac{3}{10}$
7 $\frac{7}{5} = 1\frac{2}{5}$
8 $\frac{11}{10} = 1\frac{1}{10}$
9 $\frac{17}{10} = 1\frac{7}{10}$
10 $\frac{5}{4} = 1\frac{1}{4}$
11 $\frac{7}{4} = 1\frac{3}{4}$
12 $\frac{18}{10} = 1\frac{8}{10}$
13 $\frac{11}{5} = 2\frac{1}{5}$
14 $\frac{21}{10} = 2\frac{1}{10}$
15 $\frac{17}{8} = 2\frac{1}{8}$

SET 4

1 1525
2 45
3 $\frac{1}{8}$
4 102
5 6
6 No
7 30°
8 True
9 75 L
10 2350
11 77.7
12 1 026 231
13 7100
14 1
15 Rectangular prism
16 40

Space

1

Top Front Side

2

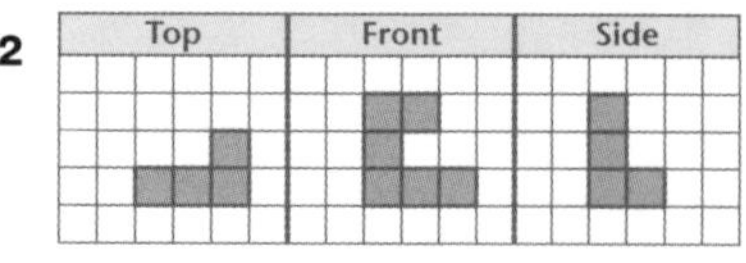

Statistics and Probability

1 0.2
2 0.2
3 0.1
4 0.4
5 0.1

UNIT 18 Number and Algebra

SET 1

1 48
2 63
3 45
4 ÷
5 6
6 +
7 Yes
8 22
9 30
10 1, 2, 3, 6, 9, 18
11 3745
12 4
13 210
14 701
15 80
16 $84

SET 2

1 19.44
2 7.73
3 11.91
4 15.82
5 8.45
6 $211.49
7 $546.74
8 18.026
9 28.791
10 **A** $333.85
B $753.85
C $877.90
11 **A** $211.25
B $268.05
C $495.11

SET 3

1 75
2 57
3 51
4 39
5 27
6 10
7 11
8 12
9 13
10 14

SET 4

1 360
2 $50
3 $9
4 60°
5 54
6 900 000
7 8
8 90 km/h
9 32
10 6
11 6 000 000
12 2257
13 $140
14 1
15 1.4 m

Measurement and Space

1 **a** 100° **b** 80° **c** 100° **d** 80°
2 **e** 120° **f** 60° **g** 120° **h** 60°

Measurement

1

Volume	100 cm³	500 cm³	2500 cm³	3000 cm³	750 cm³	90 cm³	2000 cm³	2515 cm³	1000 cm³	2900 cm³
Capacity	100 mL	500 mL	2500 mL	3 litres	750 mL	90 mL	2 litres	2515 mL	1 L	2900 mL

2 125 mL **3** 1000 cm³

UNIT 19 Number and Algebra

SET 1

1 56
2 59
3 4000
4 66
5 ÷
6 12
7 60
8 9
9 57 631
10 $4.80
11 700
12 Yes
13 No
14 5
15 $0.65

SET 2

1 651
2 6909
3 273 r 4
4 7777
5 8181 r 1
6 9870
7 11 054 r 1
8 9876 r 4
9 1500
10 $300 each
11 261 mL each
12 15
13 125
14 765
15 910
16 4
17 5

SET 3

1 $12, $48
2 $16, $64
3 $10, $40
4 $7, $28
5 20 children
6 $24
7 7 L
8 30 trees
9 30 goals
10 $16
11 25%
12 40%
13 50%
14 20%

SET 4

1 11:00 pm
2 103°
3 90
4 25% of 1000
5 0.09
6 159 000
7 30
8 1, 4, 9, 16, 25, 36, 49
9 About 6000
10 $37.10
11 248 008
12 $150
13 $\frac{6}{25}$
14 510 mm
15 $1\frac{5}{8}$
16 288

Measurement and Space

Possible dimensions:

10 cm × 2 cm
9 cm × 3 cm
8 cm × 4 cm
7 cm × 5 cm
6 cm × 6 cm

Measurement

1 10
2 20
3 4
4 100
5 50

UNIT 20 Number and Algebra

SET 1

1 56
2 180
3 5
4 +
5 11
6 1642
7 50
8 1, 3, 7, 21
9 $6.25
10 12
11 2032
12 0.06
13 ÷
14 16 025
15 71

SET 2

1 9
2 7
3 9
4 8
5 8
6 60
7 9
8 12
9 40
10 70
11 70
12 90
13 7 yrs
14 5 yrs

SET 3

1

3^2
3 × 3 = 9

2
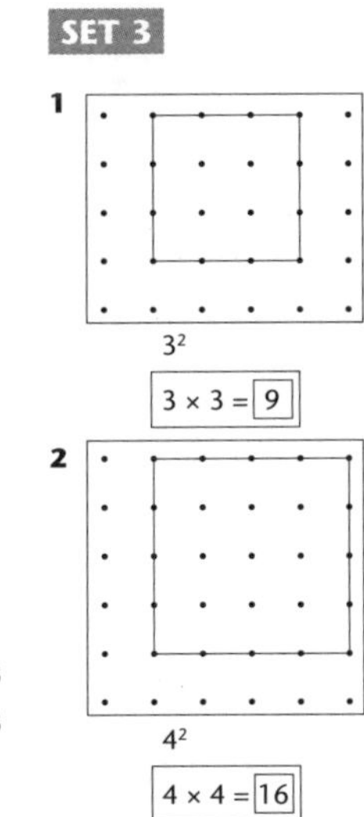
4^2
4 × 4 = 16

3
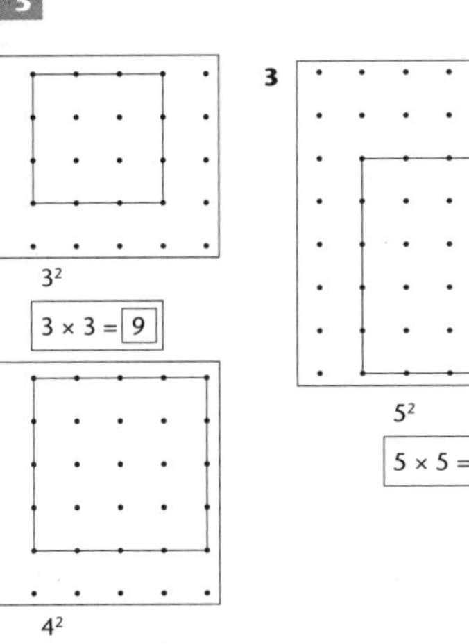
5^2
5 × 5 = 25

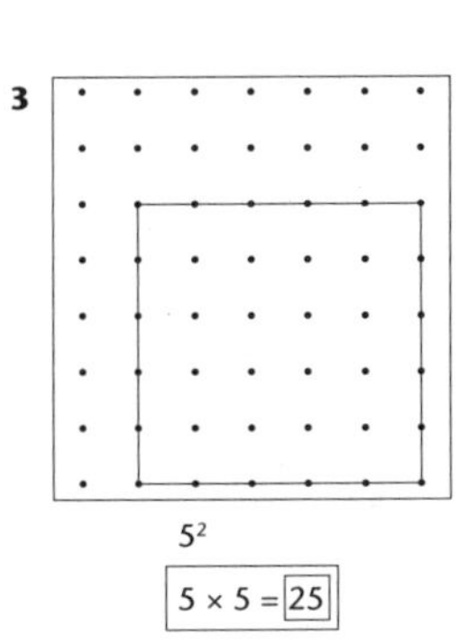

SET 4

1 85
2 $1\frac{27}{100}$
3 $2.20
4 $31.20
5 202
6 288
7 411 000
8 36
9 8
10 18
11 $34
12 Hands on.

Space

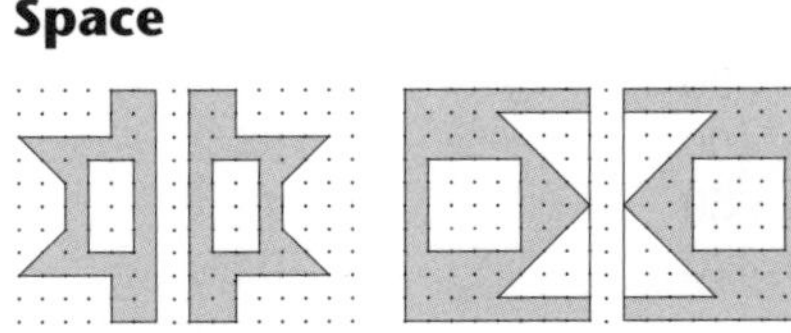

Statistics and Probability

1 True
2 True
3 True
4 True
5 True

UNIT 21 Number and Algebra

SET 1

1 32
2 10:45
3 250
4 9
5 63
6 951
7 90°C
8 14.7
9 5.9
10 172
11 30
12 81
13 67 km
14 37
15 $3.50

SET 2

1 16 380
2 7140
3 32 302
4 17 898
5 28 386
6 39 780
7 19 710
8 7683
9 42 884

10
```
    421
  ×  36
 ------
  2 426
 12 630
 ------
 15 056
```

11
```
    649
  ×  28
 ------
  5 197
 12 880
 ------
 18 077
```

SET 3

1 $\frac{17}{10}$, $1\frac{7}{10}$
2 $\frac{5}{4}$, $1\frac{1}{4}$
3 $\frac{7}{4}$, $1\frac{3}{4}$
4 $\frac{18}{10}$, $1\frac{8}{10}$
5 $\frac{11}{5}$, $2\frac{1}{5}$
6 $\frac{1}{3}$
7 $2\frac{1}{2}$
8 $3\frac{1}{4}$
9 $4\frac{7}{8}$
10 $3\frac{2}{3}$
11 $6\frac{2}{3}$
12 $3\frac{4}{8}$ or $3\frac{1}{2}$
13 $2\frac{3}{5}$

SET 4

1 19 × 4 = 76
2 $84.20
3 390
4 $13.50
5 About 12 000
6 121.67
7 $1.30
8 $1\frac{3}{10}$, 137%, 1.4, 1.45
9 54
10 827 000
11 1.27
12 10 800
13 50
14 131
15 72

Statistics and Probability

Total: 35, 10, 120

1 Lane
2 Vanda
3 Yes
4 No
5 Yes

Measurement

1

2

3

4

Answers

UNIT 22 Number and Algebra

SET 1

1 4
2 290
3 6511
4 27
5 6.9
6 135
7 $1.05
8 63
9 9.6
10 7
11 0.9
12 1
13 63 m
14 96
15 30

SET 2

1

Hexagons	1	2	3	4	5	6
Sides	6	12	18	24	30	36

2

Triangles	1	2	3	4	5	6
Sides	3	6	9	12	15	18

3

Octagons	1	2	3	4	5	6
Sides	8	16	24	32	40	48

4

Pentagons	1	2	3	4	5	6
Sides	5	10	15	20	25	30

SET 3

1 $\frac{2}{10}$
2 $\frac{3}{30}$
3 $\frac{4}{24}$
4 $\frac{6}{15}$
5 $\frac{15}{20}$
6 $\frac{8}{12}$
7 $\frac{2}{5}$
8 $\frac{3}{5}$
9 $\frac{2}{3}$
10 $\frac{2}{3}$
11 $\frac{3}{5}$

SET 4

1 0927
2 3750
3 23, 29, 31
4 $2.15
5 $\frac{3}{4}$
6 540
7 55 217
8 16
9 5 cm
10 9
11 480 cm^2
12 $224
13 Cylinder
14 $9.40
15 90

Patterns and Algebra

1 2

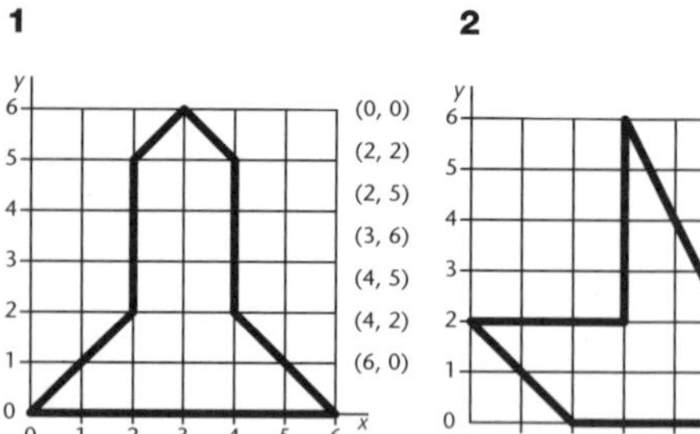

Statistics and Probability

1 True
2 True
3 True
4 False
5 True
6 True

UNIT 23 Number and Algebra

SET 1

1 30
2 500
3 400
4 950
5 420
6 ÷
7 ×
8 33
9 1, 2, 3, 5, 6, 10, 15, 30
10 No
11 23 260
12 297
13 2920
14 4
15 20

SET 2

1 52.8
2 112
3 124.2
4 71.6
5 70.56
6 206.88
7 641.1
8 1344.42
9 $1219.10
10 8.75
11 56.56

SET 3

1 $\frac{4}{10}$
2 $\frac{6}{8}$
3 $\frac{2}{6}$
4 $\frac{4}{8}$
5 $\frac{4}{6}$
6 $\frac{4}{5}$
7 $\frac{1}{4}$
8 $\frac{2}{3}$
9 $\frac{3}{4}$
10 $\frac{1}{2}$
11 F
12 T
13 T
14 F

SET 4

1 2215
2 3.19, $3\frac{1}{5}$, 3.3, $3\frac{1}{3}$
3 $300
4 $\frac{1}{5}$
5 1111
6 $33.50
7 108
8 45
9 $60
10 20:27
11 6.5°C
12 6:45 am
13 1296
14 6
15 26.33 m

Space

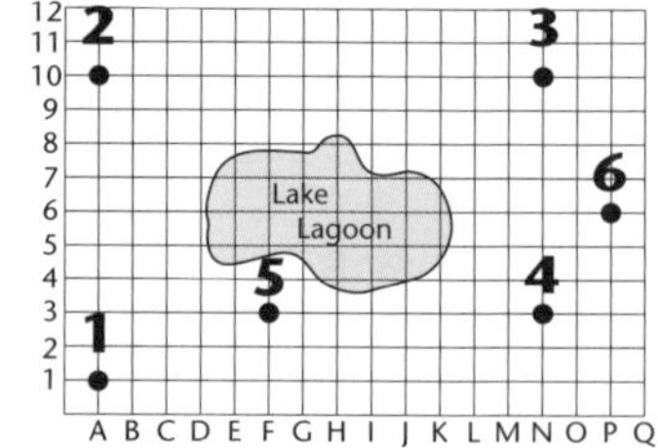

7 90 km
8 130 km
9 70 km

Measurement and Space

1

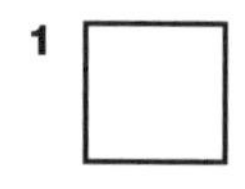

square rhombus trapezium rectangle

2 Square, rectangle and rhombus

UNIT 24 Number and Algebra

SET 1

1 28
2 Yes
3 ×
4 190
5 900
6 8
7 546
8 1, 3, 9, 27
9 $815
10 8500
11 9
12 8
13 52
14 80 000
15 450 cm
16 75

SET 2

1 452
2 258.9
3 14.5
4 86.5
5 1458
6 15.8
7 15.8
8 258.9
9 452
10 452
11 1458
12 12.5
13 86.5

NEW MILLENNIUM

SET 3

1 $\frac{3}{4}$
2 $\frac{3}{8}$
3 $\frac{3}{10}$
4 $\frac{6}{10}$
5 $\frac{4}{8}$
6 $4\frac{4}{10}$
7 $6\frac{7}{10}$
8 $7\frac{5}{8}$
9 $\frac{7}{10}$
10 $\frac{6}{10}$
11 $\frac{4}{10}$
12 $\frac{3}{8}$
13 $\frac{4}{10}$
14 $7\frac{1}{4}$
15 $7\frac{3}{8}$
16 $4\frac{4}{10}$ or $4\frac{2}{5}$

SET 4

1 4.22
2 0.35
3 4
4 141
5 $\frac{9}{100}$
6 $\frac{98}{100}$
7 45
8 $\frac{70}{10}$
9 8
10 236
11 70°
12 1 hr 40 min
13 2 hr 35 min
14 2 hr 35 min
15 1 hr 15 min

Space

1 110°
2 80°
3 50°
4 270°
5 240°
6 300°

Measurement

	Shape	L	W	H	Volume
1	A	4	2	3	24 cm^3
2	B	5	3	4	60 cm^3
3	C	6	3	4	72 cm^3
4	D	7	4	5	140 cm^3
5	E	11	2	2	44 cm^3

UNIT 25 Number and Algebra

SET 1

1 +
2 63
3 $13
4 $\frac{7}{10}$
5 30
6 –
7 ÷
8 3
9 0.6
10 30
11 No
12 No
13 235 321
14 30
15 $26

SET 2

1 $293 295
2 $916 419
3 $309 355

SET 3

1 70, 110, 260, 350, 200
2 800, 900, 1000, 1270, 3160
3 800, 1600, 2300, 10 000, 11 800
4 6000, 15 000, 50 000, 100 000, 225 000
5 $1750
6 $2380
7 $3340

SET 4

1 52
2 199
3 120
4 115
5 50
6 $\frac{7}{10}$
7 30
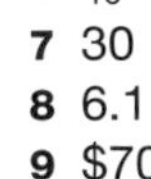
8 6.1
9 $70
10 $39.60
11 1254
12 19 × 5 = 95 or 20 × 5 = 100
13 $80
14 98, 75, 52
15 1, 2, 3, 6, 7, 14, 21, 42
16 107
17 36 cm

Space

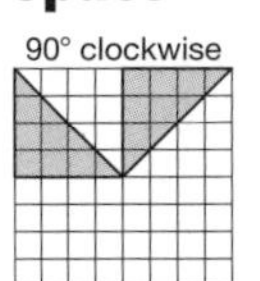

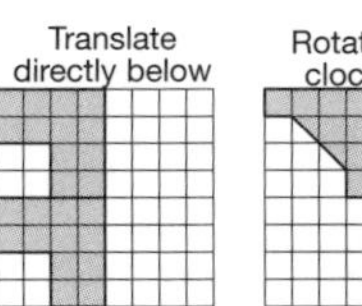

Statistics and Probability

Vanilla: 30
Chocolate: 50
Strawberry: 45
Mango: 10
Rainbow: 15

UNIT 26 Number and Algebra

SET 1

1 $5.80
2 1, 3, 5, 9, 15, 45
3 2.4
4 $\frac{9}{10}$
5 108
6 $1.20
7 13.6
8 No
9 9
10 60
11 42
12 9 r 5
13 10:06
14 19
15 9

SET 2

1 89 794
2 270 470
3 112 995
4 197 737
5 220 677
6 60 930
7 $23 598
8 24 960 g or 24.96 kg

SET 3

1 8
2 –2
3 2
4 3
5 –4
6 –2
7 –4
8 3
9 –11
10 –2
11 –18 + 7 = –2
12 $54 – $20 – $45 = –$11
13 25 – 10 – 10 + 15 = 20

SET 4

1 60 cm
2 350 m
3 60 m
4 $119.94
5 762
6 21 min
7 $1\frac{4}{5}$ or $1\frac{8}{10}$
8 75 L
9 $90
10 $8.91
11 32
12 20%
13 8
14 2.4 L
15 945 mL

Number and Algebra

1 3590, 3600, 4000
2 1060, 1100, 1000
3 3530, 3500, 4000
4 870, 900, 1000
5 9320, 9300, 9000
6 6670, 6700, 7000
7 2000 + 4000 = 6000, 6054
8 9000 – 3000 = 6000, 5916
9 6000 + 4000 = 10 000, 10 039
10 4000 + 2000 = 6000, 6004
11 8000 + 1000 = 9000, 8986
12 6000 – 4000 2000, 1932

Statistics and Probability

Hands on.

UNIT 27 Number and Algebra

SET 1

1 66
2 54
3 56
4 20
5 11:10 am
6 51
7 $72
8 1, 3, 7, 21
9 25
10 20
11 1050
12 100
13 28°
14 $1.22
15 $7.50

SET 2

1 $5\frac{1}{3}$
2 $6\frac{1}{4}$
3 $51\frac{1}{5}$
4 $85\frac{3}{4}$
5 $247\frac{1}{3}$
6 $86\frac{1}{4}$
7 $59\frac{2}{5}$
8 $131\frac{1}{6}$
9 $322\frac{2}{7}$
10 $433\frac{1}{2}$
11 $733\frac{1}{9}$
12 $359\frac{3}{10}$
13 83.25
14 198.5
15 126.25
16 186.75
17 389.7
18 267.9
19 73.2
20 1153.4
21 1075.6

SET 3

1 $1\frac{1}{4}$
2 $1\frac{1}{5}$
3 $1\frac{1}{2}$
4 $1\frac{1}{3}$

SET 4

1 20.3 m
2 $\frac{91}{100}$
3 9
4 100°
5 $8\frac{3}{5}$
6 50 000
7 $28.90
8 Various responses, e.g. 7 days × 24 hours × 60 minutes = number of minutes per week

Measurement

1 8
2 16
3 4

Space

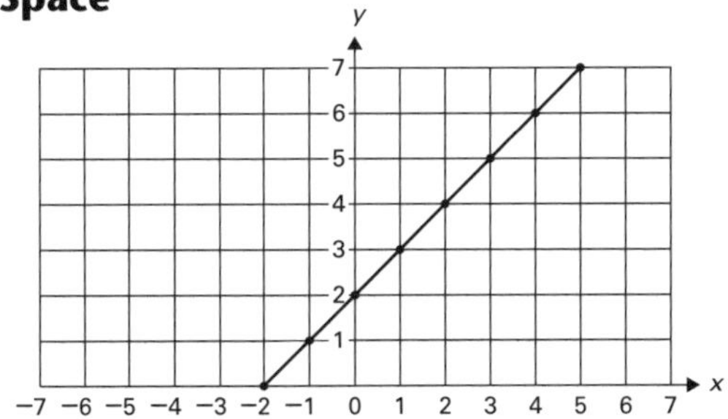

Answers

UNIT 28 Number and Algebra

SET 1

1 +
2 81
3 –
4 42
5 ÷
6 13
7 ×
8 1745
9 7526
10 $7.50
11 $7
12 8
13 15
14 6
15 35

SET 2

1 $10.80
2 $5.00
3 $7.00
4 $4.20
5 $27
6 $10.50
7 $10.50
8 $3.75
9 $2.70
10 $27.45
11 $23
12 $22.55
13 9 kg
14 8 kg

SET 3

1 35°C
2 –5°C
3 5°C
4 15°C
5 5°C
6 10°C
7 –15°C
8 –15°C
9 –10°C
10 20°C
11 5°C
12 0°C

SET 4

1 86 000
2 0.09
3 85
4 10:24
5 0.85
6 32
7 $405
8 8.06
9 7 r 6
10 25 cm^2
11 0110
12 3.92
13 720
14 $225
15 3.7, 375%, 3.79, $3\frac{8}{10}$
16 $81.63

Measurement

Peta's timeline

Number and Algebra

1 24
2 6
3 9
4 18
5 28
6 30
7 45
8 20
9 35

UNIT 29 Number and Algebra

SET 1

1 980
2 160
3 $16.50
4 9.6
5 6.5
6 250
7 51
8 $2.60
9 75
10 8 r 7
11 $\frac{50}{100}$ or $\frac{5}{10}$ or $\frac{1}{2}$
12 250
13 $100
14 250
15 120

SET 2

1 $399, $15 162
2 $250, $7250
3 $180, $7200
4 $79, $4977
5 $2.50, $247.50
6 $3.80, $361
7 $877
8 $186.30
9 $829
10 $402.50

SET 3

1 $3
2 4 matches
3 6 pens
4 4 dogs
5 10 fish
6 $15
7 12 sheep
8 12 pencils
9 150
10 180
11 120
12 100
13 60
14 150

SET 4

1 11.71
2 5
3 0.09
4 $\frac{3}{8}$
5 Rhombus
6 321
7 12
8 32 cm
9 120°
10 5 670 000
11 18 090 000
12 3 240 000

Measurement and Space

Model	Length cm	Width cm	Height cm	Volume cm^3
A	3	2	3	18
B	4	3	2	24
C	3	3	3	27

Number and Algebra

1 16 litres
2 288 kg
3 $105

UNIT 30 Number and Algebra

SET 1

1 1000
2 910
3 41
4 6
5 350
6 ×
7 +
8 4 r 1
9 12
10 60
11 220
12 174
13 100
14 5
15 2350

SET 2

1 15.9
2 18.9
3 21.9
4 24.9
5 27.9
6 1.2
7 1.1
8 1
9 0.9
10 0.8

SET 3

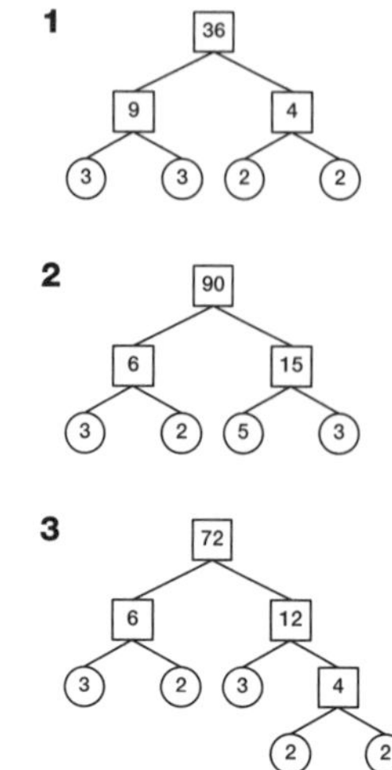

SET 4

1 585
2 $24
3 74
4 $650
5 70 × 9 = 630
6 27
7 $\frac{3}{10}$
8 7.02
9 2328
10 96 r 6
11 269
12 $2.70
13 90 m^3
14 2 000 000
15 2.9
16 $210

Measurement

Hands on.

Statistics and Probability

1 10
2 16
3 28
4 28
5 8
6 16

UNIT 31 Number and Algebra

SET 1

1 410
2 ÷
3 60
4 200
5 $1.65
6 7 r 1
7 20 231
8 900 000
9 No
10 23, 29
11 1, 3, 7, 21
12 0.23
13 $\frac{1}{4}$
14 3745
15 4500

SET 2

1 2.31, 23.1, 231
2 43.8, 438, 4380
3 6.43, 64.3, 643
4 18.7, 187, 1870
5 156, 1560, 15 600
6 356.8, 35.68, 3.568
7 429.5, 42.95, 4.295
8 23.52, 2.352, 0.2352
9 6.89, 0.689, 0.0689
10 1.95, 0.195, 0.0195
11 True
12 True
13 True
14 True
15 False
16 True

SET 3

1 56, 48, 40, 32, 24, 16, 8, 0
2 18, 36, 54, 72, 90, 108, 126
3 110, 111, 112, 113, 114, 115, 116
4 100, 90, 80, 70, 60, 50, 40
5 Hands on.

SET 4

1 680
2 21
3 9.1
4 341 000
5 0.5 ha or 5000 m^2
6 6
7 49 cm^2
8 $\frac{26}{100}$
9 100 000
10 197
11 $480
12 50
13 Five past four
14 93 mm
15 3.7 km
16 24

Measurement and Shape

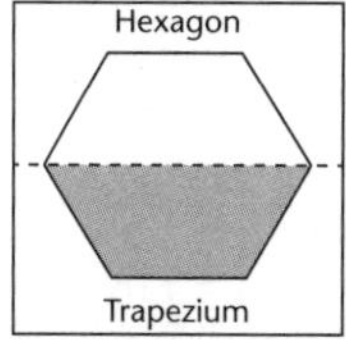

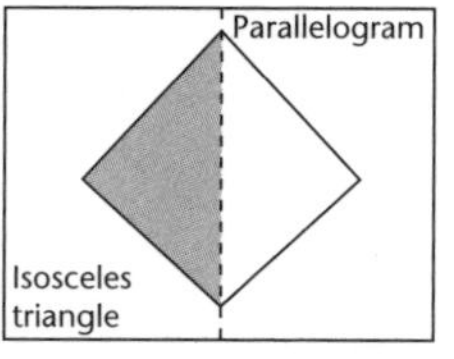

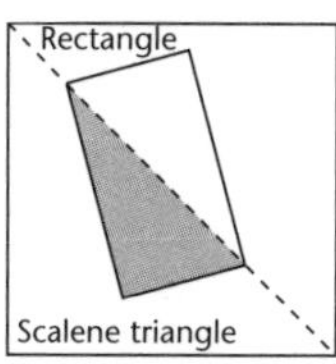

Measurement

1 647 km
2 755 km
3 766 km
4 741 km
5 1402 km

UNIT 32 Number and Algebra

SET 1

1 81
2 $1.70
3 400
4 27
5 6 r 4
6 $2800
7 $13
8 65%
9 112
10 9.83
11 3.2
12 74
13 10 000
14 0.6
15 112

SET 2

1

	×10	×100
0.7	7	70
0.6	6	60
0.3	3	30
3.5	35	350
4.8	48	480

2

	×10	×100
2.6	26	260
5.8	58	580
3.12	31.2	312
4.08	40.8	408
9.17	91.7	917

3

	×10	×100	×1000
28.52	285.2	2852	28 520
15.37	153.7	1537	15 370
18.24	182.4	1824	18 240
15.163	151.63	1516.3	15 163
36.202	362.02	3620.2	36 202
72.374	723.74	7237.4	72 374
51.727	517.27	5172.7	51 727

SET 3

1 5
2 6
3 6
4 3
5 6
6 14
7 16
8 12
9 25
10 30
11 False
12 False
13 True
14 True
15 False
16 Book C

SET 4

1 True
2 35 240
3 20
4 3
5 25 000
6 6000
7 37°C
8 5 302 000
9 225 cm^2
10 About 120 000
11 $82.80
12 1.655 km
13 4350
14 4680
15 500
16 546
17 70
18 172 cm

Space

1 A = (4,4)
2 B = (2,3)
3 C = (1,1)
4 D= (4,1)
5 E = (–4,3)
6 F = (–2,1)
7 G = (–3,–2)
8 H = (2,–2)

Measurement

1 395
2 296
3 245
4 98 885 kg
5 50 g
6 940 g
7 5 kg
8 450 g
9 505 g
10 450 g
11 3.25 kg

UNIT 33 Number and Algebra

SET 1

1 5 r 1
2 2
3 3400
4 ×
5 +
6 70
7 15
8 3 r 3
9 37
10 50
11 1500
12 $23
13 37 726
14 $12
15 $45
16 5 (7, 11, 13, 17, 19)

SET 2

1

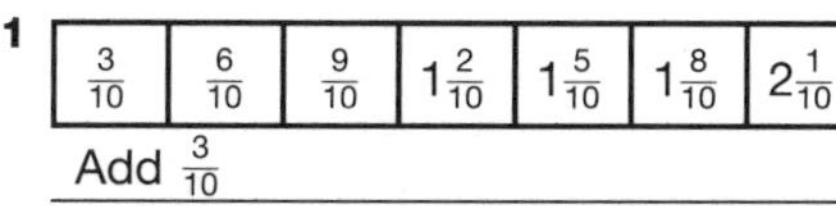

$\frac{3}{10}$	$\frac{6}{10}$	$\frac{9}{10}$	$1\frac{2}{10}$	$1\frac{5}{10}$	$1\frac{8}{10}$	$2\frac{1}{10}$

Add $\frac{3}{10}$

2

$4\frac{7}{10}$	$4\frac{3}{10}$	$3\frac{9}{10}$	$3\frac{5}{10}$	$3\frac{1}{10}$	$2\frac{7}{10}$	$2\frac{3}{10}$

Subtract $\frac{4}{10}$

3

1.7	2	2.3	2.6	2.9	3.2	3.5

Add 0.3

4

0.8	1.6	2.4	3.2	4.0	4.8	5.6

Add 0.8

5

5.51	5.54	5.57	5.60	5.63	5.66	5.69

Add 0.03

6

$2\frac{3}{8}$	$2\frac{6}{8}$	$3\frac{1}{8}$	$3\frac{4}{8}$	$3\frac{7}{8}$	$4\frac{2}{8}$	$4\frac{5}{8}$

Add $\frac{3}{8}$

SET 3

1 6
2 5
3 18
4 2
5 2
6 25
7 5
8 4
9 ($500 – $125) × 52 = $19 500

SET 4

1 8750
2 $42.75
3 155
4 True
5 12
6 20%
7 100°
8 87 km/h
9 1620 cm
10 120
11 326
12 0.1, 0,11, 1.0, 11.1
13 $200

Statistics and Probability

1 39
2 58
3 74
4 80
5 118
6 42
7 110
8 91
9 300

Measurement

1 180 m^2
2 1080 m^3
3 $324 000

Answers

UNIT 34 Number and Algebra

SET 1

1 500
2 ÷
3 270
4 +
5 600
6 92
7 False
8 76 000
9 25
10 78
11 True
12 560
13 60
14 False
15 $1.80

SET 2

1 $12, $15, $18, $21
2 6 L, 7.5 L, 9 L, 10.5 L
3

Pentagons	1	2	3	4	5
Sides	5	10	15	20	25

SET 3

1 $60
2 $96
3 $64
4 $70
5 18 km

SET 4

1 754
2 15.5
3 48
4 94 km/h
5 33%
6 0
7 $126
8 $\frac{3}{4}$
9 120°
10 Various solutions, such as:
$10 \times 6 \times 1$
$10 \times 2 \times 3$
$5 \times 6 \times 2$
$5 \times 3 \times 4$
$12 \times 5 \times 1$
$20 \times 1 \times 3$

Statistics and Probability

1 8°C 2 8°C 3 8°C
4 0°C 5 0°C

Measurement and Space

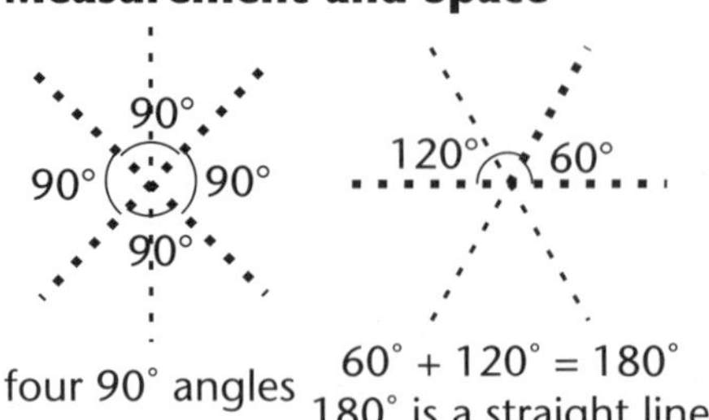

UNIT 35 Number and Algebra

SET 1

1 58
2 $20
3 20
4 ÷
5 100
6 –
7 214
8 72
9 0.5 L
10 No
11 12:07
12 60 000
13 39 241
14 72
15 366
16 1 461

SET 2

1 8
2 12
3 48
4 27
5 5
6 87
7 16
8 7
9 3
10 93
11 12
12 7
13 48 ÷ 6 = 2 × 4
or
48 ÷ 6 = 4 × 2
or
48 ÷ 8 = 2 × 3
or
48 ÷ 8 = 3 × 2
or
48 ÷ 4 = 6 × 2

SET 3

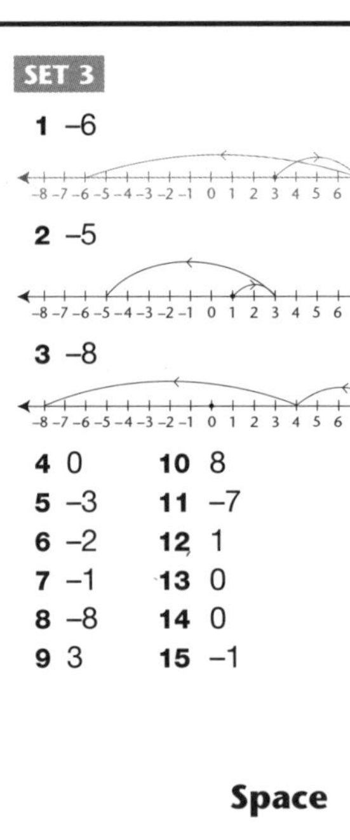
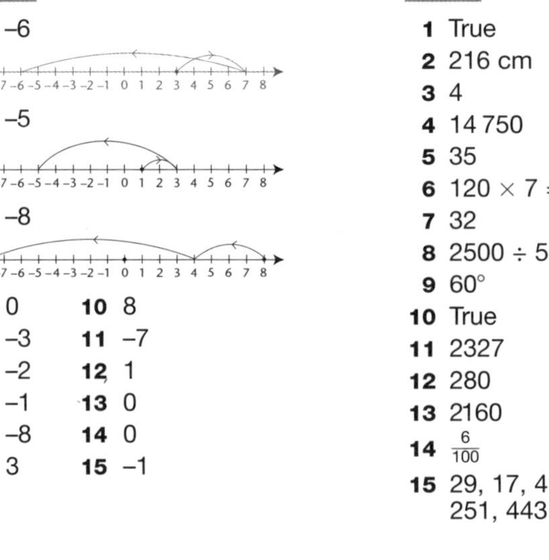

1 –6
2 –5
3 –8
4 0
5 –3
6 –2
7 –1
8 –8
9 3
10 8
11 –7
12 1
13 0
14 0
15 –1

SET 4

1 True
2 216 cm
3 4
4 14 750
5 35
6 $120 \times 7 = 840$
7 32
8 $2500 \div 5 = 500$
9 60°
10 True
11 2327
12 280
13 2160
14 $\frac{6}{100}$
15 29, 17, 41, 131, 251, 443

Statistics and Probability

Age	Tally	Frequency
14	|||	3
15	卌	5
16	卌 ||	7
17	卌 卌 |||	13
18	卌 卌 卌	15
19	卌 ||	7

Space

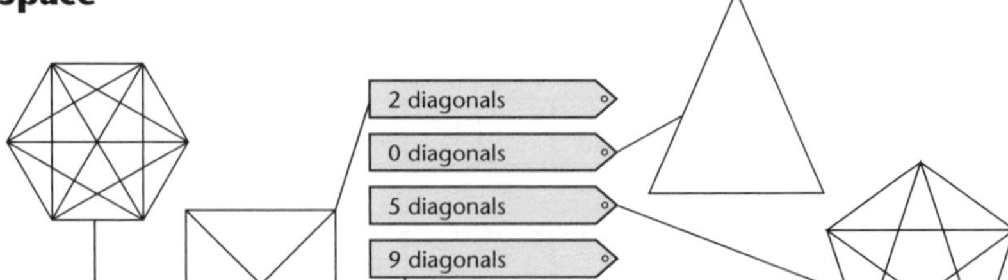